Co·GENESIS

Also by Randolph R. Croxton:

A Convergence of Two Minds

Co·GENESIS

THE RNA MIND: PERCEPTION & CONSCIOUSNESS

Randolph R. Croxton

PALUSTRIS PRESS

Palustris Press
44-02 23rd Street, Studio 412
Long Island City, NY 11101

| | | | | | | | | |

Co·GENESIS

is a theory of RNA primacy, perception, and consciousness in sexually-reproducing
life forms. Separate ancestral lines of male and female life experience are asserted here
as the binary source of near-infinite variation in individual perception and the larger
social order of the species—essential to their resilience. Contemporary research, ranging
across the fields of the life sciences: neuroscience, psychiatry, and evolutionary theory, are
integrated in support of the argument. Importantly, all the research teams are credited
directly in the text and in greater detail in the *Notes*.

In order to make the work more accessible, there are 80 *Figures*
(artwork, photos, analysis, and digital images), many with page-opposite explanations.
Sources and crediting for all *Figures* are contained in *Image Credits* at the end of the
book. Licenses have been acquired for individual images within
5 of the 80 *Figures*: numbers *12, 37, 62, 63*, and *72*, which are
individually called out in *Image Credits*.

Cover design: An imagined view of the Garden of Eden is
the author's interpretation from photographs of two-dimensional artwork.
The surrounding garden is from Henri Rousseau (1844-1910), a sampling of
backgrounds from his *Virgin Forest with Sunset* (CCO 1.0)
Universal Public Domain Dedication
and the two central images, of Adam and Eve are
by Albrecht Dürer (1507): collection of the Prado Museum,
Madrid, Spain, Getty Images, see *Image Credits*.

| | | | | | | | | |

Book Graphic Design by Jean Hahn
Manufactured in the United States of America
Library of Congress Control Number: 2022922164
ISBN 978-0-9961176-1-6

Dedicated to Fran Drummond, my wife,
partner, literary critic and loving champion of family, friends, and
the pleasures of life.

My thanks to Professor Vernon F. Shogren of
the School of Design, NCSU, Raleigh, NC who
taught his students the hard pathway to an original thought.

CONTENTS

FOREWORD

This book is about you—why you have a unique point of view and why no one else has (or ever will have) your perception of the world. Such near-infinite variability of perception/behavior creates a resilient human species: a complex story that is told simply in the pages of this book.

It is difficult to imagine that a misconception about the two molecules that carry life's information, DNA and RNA, could result in a misunderstanding of who we are as humans. And yet the belief that our DNA is producing a "normal" female or male mind with a normal level of empathy and caring, a normal level of distrust and aggression, or a normal attraction between the sexes is the ultimate failure to grasp the nature of humankind. While the physical body evolves to a steady state of fitness through natural selection and the mixing of male and female genetic material consistent with our DNA-centered understanding, our minds vary dramatically across a wide spectrum of perception/behavior.

Survival in this world requires contingent decisions: trust or distrust, fight or flight, forgiveness or revenge…it just depends. Life is therefore an indeterminate challenge in advance of an unknowable future. Charles Darwin believed that the inheritance of life experiences from previous generations was necessary to impart the high degree of variation (not normalcy) required for natural selection to act on when shaping a species to its environment; we will embrace Darwin's opinion as we consider the pattern of diversity (social order) of all minds.

The secret to the variation of perception-behavior asserted here is that we inherit two minds from our parents: a left hemisphere of ancestral male life experience and a right hemisphere of ancestral female life experience. Each individual is a unique interconnection of these sourcebooks—our perception of the world, our personality. Only in Humans do these minds converse; our silent inner voice of *deliberative consciousness*: the basis for the ±7.8 billion unique Human minds on Earth today and easily the basis for 7,800 billion more for 1,000 near-earths to come—a central premise of Co·GENESIS.

A foundational bit of knowledge in the fields of Evolution and Biology is that DNA carries information from one generation to the next in long helical strands that include chapters of heritable information called genes that travel within egg and sperm (gametes). These strands contain the individual Code of Life, the blueprint for body, regulation, and reproduction written in DNA molecular form. DNA has been at the center of evolutionary theory since the discovery of the double helix in 1953. For this and many other reasons, it is easy to conclude that DNA controls all life processes. To the contrary, DNA is the form of *inactive Code,* a reference copy, that can be used to create *active Code* (RNA molecules) in a process called *transcription*; engineering the change is no other than RNA itself. The Code of Life in the form of DNA remains locked in the nucleus of the cells over a lifetime.[1]

The DNA helix is the perfect molecular structure for information storage— a stable geometry that can withstand the gymnastics of cellular division and the transits of the reproductive cycle. By contrast, RNA active molecules travel throughout the liquid rivers of the body from the inner sanctum of the cell, the nucleus, through the walls of the nucleus to the outer domains of the cell, and onward (extracellular) throughout the body and reproductive pathways to direct the dynamic of life. Upon a signal from RNA the cell and its DNA divide (the source of more RNA) and through a process of *apoptosis* (cellular death) RNA removes cells. Cells with their inactive DNA Codes abide in a constant ocean of active RNA; RNA, not DNA, is in charge. For this reason I will refer to RNA, the active form of information, when describing the biological dynamic of life.

As made clear in this book, the super-power of RNA is its ability to maintain two separate lines of informational descent: RNA-imprinted male genes and RNA-imprinted female genes, which interact in an oppositional yet complementary fashion (a binary). A unique blending of these streams of ancestral knowledge inform the hemispheres of the brain in the process of birth—updating and diversifying Minds for a resilient species. These RNA pairings act much as the 0's and 1's in the binary code of a computer. Although the RNA couplets comprise less than 1% of the body's genes, they author, maintain, and edit the Code of Life stored in DNA. DNA does not decide which information is to be used any more than a librarian decides which books you read. Instead, the decision-maker is the RNA MIND; the male and female duo that acts as the patron, first in building the DNA library, and then expressing, editing, and withdrawing from it to regulate and reproduce life.

The Theory of Co·GENESIS asserts that separate RNA male and RNA female lines of *imprinted genes*[2] are passed down through the generations, evolving separately but joining in each individual to uniquely express the MIND. The left (male) hemisphere and right (female) hemisphere of the brain sit astride the *hypothalamus,* the conductor of the orchestra: the nervous system (perception/behavior) and the endocrine system (metabolism and reproduction). Interconnected, and acting as one, this is the RNA MIND–the way we see the world.

The concept of the RNA MIND represents a far deeper biological unity of humankind, male and female, than previously understood and reverses long-held beliefs about the molecular chain of command. The flow of information is not DNA-to-RNA, but the other way around in a more complex RNA-to-DNA transfer. Persistent questions of human perception and behavior are clarified in this new biological and evolutionary context:

1) Why are there identical twins but no identical minds?

2) Where do the two voices of our inner conversation come from? (Your inner voices are the silent thoughts that you are using to read these words. When we deliberate, the two viewpoints of our mind—maleness and femaleness—are in conversation).

3) Why is it that 40% of the population is consistently more aggressive and predisposed to see a world of threat (conservative), while another 40% is more empathetic, seeing a world of opportunity (liberal)?

4) Why are independents consistently 20% of the population in elections?

5) Why are psychopaths (1%) and sociopaths (3%) a small but constant proportion of the population?

While the existence of separate RNA-imprinted genes of male and female descent is well known, their role in inheritance and reproduction has long been the subject of debate and multiple hypotheses. The prevailing view is that all genes and information are DNA and that the chain of command is DNA-to-RNA. A quick history of the DNA-centered position is therefore in order.

Why are you the way you are? The answer lies in the fields of Evolution and Biology. Although reproduction and inheritance appear complicated at first glance, they are centered on two simple questions: how is information passed from one generation to the next and what is the nature of that information? Plato and other Greek philosophers thought that the Mind and Body are separate; the immortal Mind (intellect) embodying Truth and Wisdom, existing before birth and surviving death, while the Body is merely mortal, a one-time use. Charles Darwin thought information came from interactions with the environment, life experiences passed down in small particles in the bloodstream to egg and sperm and on to the next generation. However, neither Plato's separation of Mind and Body nor Darwin's above-stated concept of *Pangenesis* (1868) had much currency after 1900. In that year, the earlier studies (1856-1863) of the Augustinian friar Gregor Mendel gained widespread recognition as establishing the rules for the heredity of information. Mendel based his theories on a disciplined analysis of plants and his own deduction of hereditary genes (one from each parent). The clarity of the mathematical basis of what came to be known as *Mendelian Inheritance* and *Genetics* was irresistible to scientists.

In 1937 Theodosius Dobzhansky, a Russian-American biologist, put forth a *unified genetic model*, being a joint mathematical construct that incorporated Genetics and Darwin's Theory of Natural Selection; in 1970 he expanded that vision to include the ground-breaking Watson/Crick molecular revelations of the DNA double helix (1953). This unified framework came to be known as the "Modern Synthesis." With minor adjustments, this has been the baseline of DNA-centered concepts for over sixty years, despite growing inconsistencies (characterized as *non-Mendelian* or *non-Watson/Crick* to this day). The failure of DNA-centered theories to account for the speed and complexity of evolutionary adaptation has become evident; a reconceptualization is long overdue.

In *A Convergence of Two Minds* (Palustris Press: 2015), I first asserted that our perception/behavior is drawn from ancestral life experiences, uniquely expressed by RNA in the brain's separately evolving hemispheres: male legacy left, female legacy right. The DNA Body was seen as a direct inheritance of physical fitness (minimal variation, heritable) in contrast to RNA perception/behavior (maximum variation, not heritable). The DNA Body and the RNA Mind were envisioned as separate but equal pathways—that is no longer the case. The Theory of Co·GENESIS asserts that RNA is the first form of life on Earth and that evolution proceeds under RNA primacy.

A molecular biologist in reviewing the assertions of *Two Minds* in 2015, challenged the premise that life experiences in one generation can pass to the next because male and female DNA/RNA are randomly mixed in the first cell as it divides after fertilization. I immediately began following emerging research to counter this conventional argument. By October of 2017 research teams in France, Canada, and the United States had published findings consistent with three key working assumptions in *Two Minds*:

1) RNA as the first life form under the RNA World Hypothesis.[3]
2) A Biological basis for the synthesis of DNA from RNA.[4]
3) Evidence of RNA imprinted genes as networks of brain function, neuro-developmental processes, puberty, and sexual behavior.[5]

The primary object of my search arrived nine months later, when Judith Reichmann et al., under the leadership team of Jan Ellenberg at the European Molecular Biology Lab (EMBL) in Heidelberg, Germany, published a paper on the key mechanism of the RNA MIND[6] in *SCIENCE* (July 2018).

Utilizing an inverted light-sheet microscope of their own making, researchers discovered a "dual-spindle formation" which maintains the separation of parental genomes (the half-sets of biological information in sperm and egg cells) during the zygote-to-first-cellular-division. See *Figure 1*.

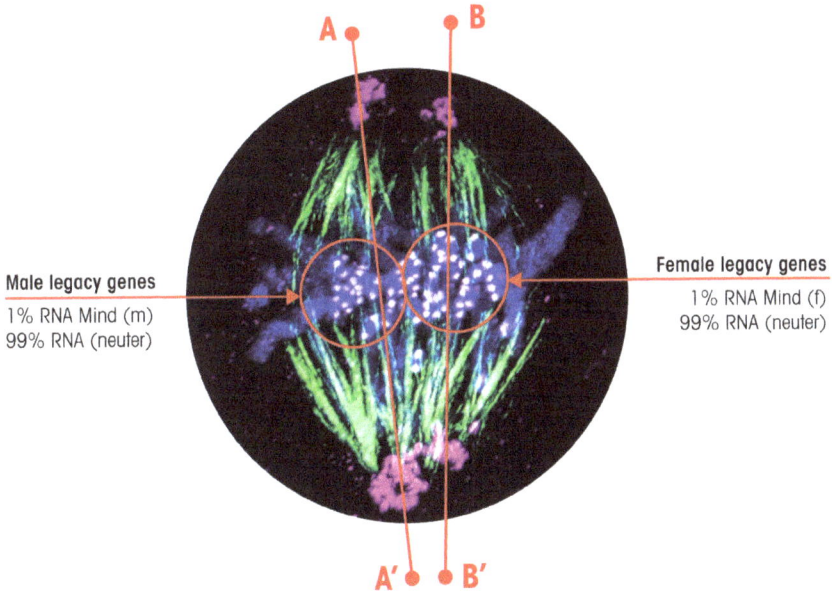

A •   • **B**

Male legacy genes
1% RNA Mind (m)
99% RNA (neuter)

Female legacy genes
1% RNA Mind (f)
99% RNA (neuter)

A' •   • **B'**

Figure 1. Male | Female Legacy Separation

Delayed genetic mixing in the first three-to-six days of transit from fertilization to the womb opens a window for separate RNA male and RNA female cells—the reset of Co·GENESIS in new life. The separated genomes appear to be in suspended animation during transit, however, RNA is reconciling and updating a new DNA Code of Life. As we will see, RNA molecules are at work as the embryo passes down the fallopian tube into the RNA information-rich ocean of the womb.

Elated by the Ellenberg/Reichmann validation after nearly three years of pursuit, I imagined that the mechanism by which the RNA Mind passes information to the embryo would be self-evident, but the opposite was true. The EMBL Team's recasting of the early developmental stage—the approximately six-to-twelve days from fertilization to implantation in the womb—posed a complex challenge. Separated gene sets and delayed merging of genomes led me to drop DNA/RNA equality. I started over. Research and finding a basis for RNA primacy took from July 2018 to July 2019, when I gave the first lecture on Co·GENESIS at the Century Club in New York City. That reconceptualization is the subject of this book.

The conclusion is that there is only one conductor of the orchestra: binary (male/female) RNA. To use a grammatical metaphor, RNA is the verb (acting) and DNA is the noun (acted upon). The RNA MIND (based on the gene sets imprinted) is the *nervous + endocrine system* coordinated by the mid-brain *hypothalamus*, which is the largest concentration of RNA-imprinted genes in the body.[7] The RNA 99% of genes that are not imprinted have their ancestral male and female genetic information mixed and are therefore unary, not binary. Although appearing to be self-replicating and autonomous, DNA is incapable of dynamic functions such as maintaining our day/night cycle and regulating the body; DNA simply serves as the cellular reference library, expressed and animated by the binary RNA MIND over a lifetime.

The Theory of Co·GENESIS reasserts Darwin's *Pangenesis* hypothesis, which envisioned *gemmules* (units of information) passing down life experiences and adaptations through sperm and egg. Pangenesis became a defining debate between the supporters of Charles Darwin vs. August Weismann, who prevailed with the argument that no modification of the body cells as a result of life experience could be passed down. This assertion, later coined the *Weismann Barrier*, became a central premise of the Modern Synthesis. In the face of the initial criticism of Pangenesis, Darwin wrote to his friend Joseph Hooker in 1868 that he would eventually be proven correct:

> "I fear Pangenesis is stillborn; Bates says he has read it twice, and is not sure he understands it…Old Sir H. Holland says he has read it twice and thinks it very tough; but believes that sooner or later 'some view akin to it' will be accepted. You will think me very self-sufficient, when I declare that I feel sure if Pangenesis is now stillborn it will, thank God, at some future time reappear, begotten by some other father, and christened by some other name."[8]

The pathways of the domains: RNA Body and RNA MIND are shown in *Figure 2* (The "Body" includes the physical Mind and Body and all unary regulatory agents). On the left are the single-line pathways of the RNA Body gene flow in green and on the right are the pathways of the RNA MIND overlaid in black. The binary pathways connect left-to-left hemisphere (male legacy sperm), and right-to-right hemisphere (female legacy egg). The RNA Body is unconscious and neuter until expressed by the RNA MIND. Metaphorically the Body of a species is an avatar, the means by which RNA can swim, walk, or fly over the Earth.

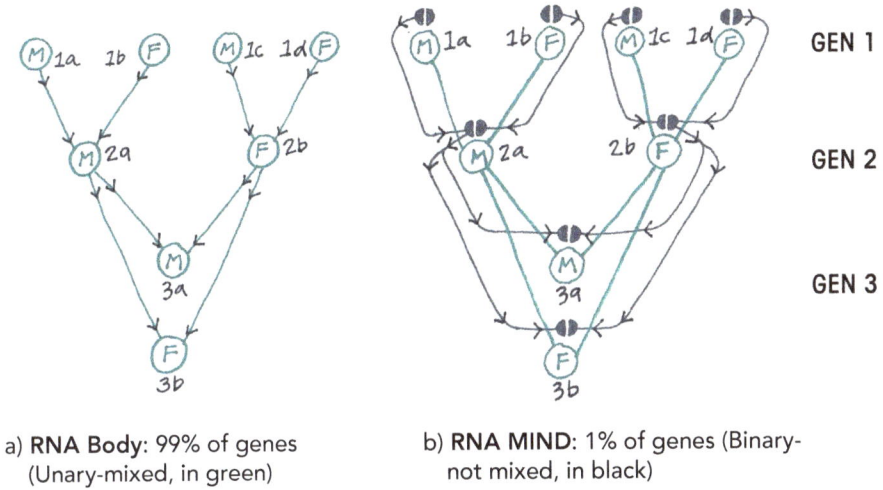

a) **RNA Body:** 99% of genes (Unary-mixed, in green)

b) **RNA MIND:** 1% of genes (Binary-not mixed, in black)

Figure 2. Three Generations of RNA Body and RNA MIND

No one can doubt the decisiveness of Golda Meier, Margaret Thatcher, or Angela Merkel in the leadership of their respective nations (*high maleness*) or the innovative/artistic sensibility of Michelangelo, Botticelli, or Leonardo da Vinci in creating the exemplars of Western Civilization (*high femaleness*). Binary modulation of emphasis and interconnection between the hemispheres, *maleness* (left) and *femaleness* (right), forms a near-infinite spectrum of individual perception/behavior across the population, ranging from the maleness extreme of near-total aggression and lack of empathy in sociopaths and psychopaths to the femaleness extreme of maximum empathy by those who sacrifice their lives for others, heroes and saints. In the formation of the Body, we are male or female but in the formation of the Mind, we are male and female. Wars and meteor strikes allow no time for a species to evolve the perception/behaviors needed for survival, therefore, in war or peace, all capabilities must be continuously present across the social order of a resilient population. No one sees the world in the same way.

Egg and sperm join to form offspring (RNA male genes inform the male cells and RNA female genes inform the female cells of the RNA MIND and RNA mixed genes inform the cells of the Body). However, the re-separation (imprinting) step required to form reproductive half-cells in the male or female fetus occurs four months later, when the developing Mind has reached the necessary degree of updating and integration for the process of *gametogenesis*. Each one of us carries two versions of the Mind: the binary version which is our bi-hemispheric brain and nervous/endocrine system for life, and the evolving unary version which is one half of the next-generation RNA MIND: our gametes (female eggs, or male sperm). See *Figure 3*, which describes the descent of the imprinted genes: two streams, father-to-son and mother-to-daughter, achieved by a process of "silencing." RNA silences the anatomically opposite parental information which is shown as "X's," creating single-sex rivers of information: male sperm (blue) on the left and female eggs (red) on the right. As a result, two separately evolving minds can combine in offspring (GEN 5) for a near-infinite number of expressions.

The two 65-million-year mammalian lines shown here date back to the end of the dinosaurs during the *K-Pg extinction event* however, imprinting goes back ±1.0 billion years to the advent of sexual reproduction. Under our current DNA-centered model, the advantage of DNA sexual reproduction is said to be diversification through the random genetic mixing of the genomes in the first cellular division. To the contrary, the advantages of RNA sexual reproduction include much more—the inheritance of parental adaptations, the overall social order of species, and the deliberative/conscious Mind of Humans. The Theory of Co·GENESIS asserts that evolution in sexual animals proceeds under RNA Primacy and the RNA biological mechanism of *Genomic Imprinting* which maintains separate male and female legacies.

Humankind is at a perilous crossroads due to the misleading simplicity and omissions of the DNA model. Scientists have confidently stated their intention to go beyond gene therapy in individuals and cut into the reproductive pathways (the Code of Life) to permanently "fix" future disease and defect. A Chinese neo-Dr. Frankenstein, used a molecular-modification technique (CRISPR-Cas9) to cut into the reproductive cells (germline) of two female embryos in 2018. He claimed to have orchestrated the birth of twin girls with immunity to their father's HIV. The Chinese Court in December of 2019 sentenced the rogue researcher He Jiankui, who admitted to multiple "off target" errors, to three years in prison.[9]

An unexpected mix of well-deserved recognition and accelerating peril has just broken like a wave across the life sciences and humanity. Jennifer Doudna and Emmanuelle Charpentier received the Nobel Prize for the above-mentioned CRISPR Technology in October 2020 and Walter Isaacson authored The *CODE BREAKER*, Simon and Schuster (March 2021) about Doudna's award-wining work.[10] I have the highest respect for Doudna, Charpentier, and Isaacson, but breaking a Code, much like the German Code in World War II, requires that you can both read and understand it.

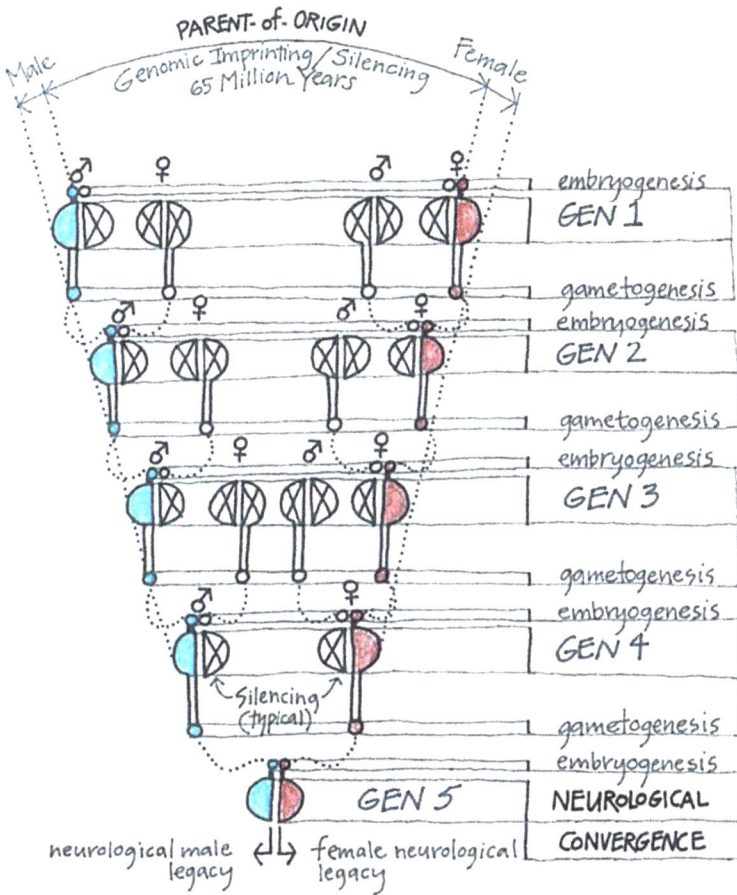

Figure 3: Mammalian Genomic Imprinting: Pathways of the RNA Mind

We understand how to modify the Code, but we don't recognize RNA primacy and its binary variation that forms the Code. Proceeding to edit the human germline at this point will bring irreparable tragedy to humankind. **Modification of human reproductive cells must stop, and an open, critically-informed, public/scientific discourse must begin.** To this end, Co·GENESIS contains eighty illustrations, easily followed image-to-image with a minimum of technical language.

New York, New York
September 2022

CHRONOLOGY OF AN IDEA

PLATO
428–348 BC
The Mind is immortal and separate from the temporal Body. The Human Body and senses are misleading; truth exists in the *rational introspections* of the Mind.

JEAN-BAPTISTE LAMARCK
1744–1829
Lamarck proposed that the Mind's adaptive information, lessons from life experiences, pass down to the next generation, a form of *transmissible memory*.

CHARLES DARWIN
1808–1882
Darwin's 1868 Hypothesis, *Pangenesis*, provided a biological basis for Lamarck's Theory. Gemmules, particles in the blood, were seen as the way for adaptive information to travel from the organs onward via sperm and egg (gametes).

AUGUST WEISMANN
1834–1914
Weismann countered Lamarck-Darwin by asserting that only *reproductive cells*, the gametes, can carry hereditary material (later called genes). Therefore, adaptive information could not flow from the *body cells* to the gametes as Darwin stated in Pangenesis. This widely-adopted separation of cells became known as the *Weismann Barrier*.

GREGORY MENDEL
1822–1884
Mendel was an Augustinian friar whose analysis of plants led to his deduction of hereditary genes (one from each parent). The validation of his work in 1900 was seen as consistent with Weismann, and established the rules for inherited characteristics, the basis of *Genetics*. Darwin and Lamarck's assertions on the inheritance of adaptation were set aside.

JAMES WATSON and FRANCIS CRICK
1928–   *1916–2004*
Watson and Crick co-authored the paper which first described a 3-D model of the double helix DNA molecule in 1953. The DNA helix of hereditary material and genes came to be seen as the Genetic Code of Life while the other informational molecule, RNA, was considered secondary. In 1960, Crick lectured on the flow of information as DNA-to-RNA and DNA-to-Protein. The current concept combining elements of Darwin, Weismann, Mendel, and Watson/Crick, is referred to as the *Modern Synthesis*.

Co·GENESIS
2022
Co·GENESIS adopts the Plato, Lamarck, and Darwin (Pangenesis) Philosophy that life experiences pass down the generations, and asserts that RNA, not DNA, directs biological function. Male and female life experiences in sexually reproducing animals are seen as informing the binary RNA MIND in offspring: [RNA $_{MALE}$ + RNA $_{FEMALE}$].
The RNA Body is expressed by the RNA MIND as male or female. RNA forms the binary Code of Life (biological intelligence). RNA regulates bodily function and form. RNA imparts the social order of perception/behavior in animals. In Humans only, the RNA MIND evolved to the overlay of *deliberative consciousness*.

CHAPTER ONE

THE BRIDGE

Perception of the world is imparted by Nature and Nurture. The first is an inborn predisposition, your personality; the second is life experience and learning. In my case, training as an architect has shaped how I observe the world and pushes me to consider how things relate, how they can be assembled as a whole: integrative thinking. My particular focus, human-centered design, led to the study of the life sciences. Human perception and behavior thus formed a bridge to the larger undertakings of this book.

Long before I began my architectural practice, an international Modernist movement combined with emerging technologies to bring about a decisive shift in architecture, changing the relationship between the built and natural environments. From the rock outcroppings and caves of earliest human habitation, to the 12th century cliff dwellings of Mesa Verde, to the soaring heights of the (1933) RCA Building centerpiece of Rockefeller Center—all were tied to natural systems: fresh air, full-spectrum natural light and the daily traverse of the sun. Rockefeller Center was formally completed in 1939, and in the same year (1939-40), the New York World's Fair opened with the first global exposition focused on the future (the motto was *The World of Tomorrow*). The three technological cornerstones of the Modernist movement were first introduced in the United States at that time: air conditioning, fluorescent lighting, and a sealed glass curtain wall building by architects Skidmore, Owings & Merrill. Suddenly there was no need for the slender buildings like the RCA Building with nearby operable windows, daylight, and natural air-flow. Buildings could have massive floor plates with engineered light levels, and cooling with minimal outside air and few occupants near a window (typically not operable). At the time, these buildings with stable temperatures in summer and winter and the clean lines of the modern aesthetic embodied the arrival of the Modern World.

However, the Oil Embargo of 1973 ignited demands for energy efficiency, which meant for the new-style buildings, cutting in half their already minimal use of outside air (fresh air that needs to be heated or cooled). The introduction of formaldehyde, petroleum solvents, and glues in interior fabrics, finishes, and furnishings further undercut the quality of indoor air.

The following era of "Sick Buildings" was symbolized by the Philadelphia Legionnaire's Disease that killed 29 in 1976. Ironically many of these post-1973 energy-efficient buildings were called "environmental" even as they were failing to address many issues of health and well-being. People's biological connection to daily and seasonal rhythms was nearly severed. Against this backdrop, I started an architectural firm in 1978.

Kirsten Childs (as co-Director of Croxton Collaborative) and I had developed a modest range of design enhancements to address these growing environmental challenges, and during the period 1985-1988, we developed an integrated human-centered health and well-being design approach which included a dimension being lost in modern architectural interiors: *circadian rhythm*. *Figure 4* depicts Leonardo da Vinci's *Vitruvian Man* standing between the sun and the moon and within our 24-hour circadian cycle. A function of planetary motion (daily rotation of the Earth on its axis), circadian is joined by *circannular rhythm*: one orbit of the Earth around the sun with Earth's tilted axis imparting the seasons of the year. All animals are biologically woven into these rhythms through an orchestration of the cycles of sleep/wake, hormones such as cortisol and melatonin, body temperature, metabolism, alertness, etc. (see *Figure 4*).

Acting on intuition as much as solid science, we designed a 3-story floor cut and planar wall (north/north-west to south/south-east) with skylights over straight-run stairs in the NYC Headquarters of the Natural Resources Defense Council (NRDC): the client who urged us on to a deeper consideration of *ecologically-informed design*.[1] Well-being, we felt, was enhanced by "reading" the daily traverse of sunlight within the space. See *Figure 5*. Having stairways in the "outdoors" and corridors open-ended to exterior views meant knowing not only "where you are" but also "when you are" in the course of the day. We never imagined that thirty years later, the Nobel Prize would go to three scientists who unraveled the RNA molecular mechanisms of this connection as being central to our health and well-being, or that molecules throughout the body (miRNAs) oscillate in tune with these rhythms. The fresh air ratio for indoor air quality and oxygen levels post-Oil Embargo had been cut to 7 cubic feet per minute per person; we increased it by 328% to 30 cfm and still beat the then-current energy code by >50%. NRDC, a national exemplar, was instrumental in raising the ASHRAE national ventilation rate in offices by 142%, to 17 cfm[2] a reversal of the downward trend in this measure of health and well being.

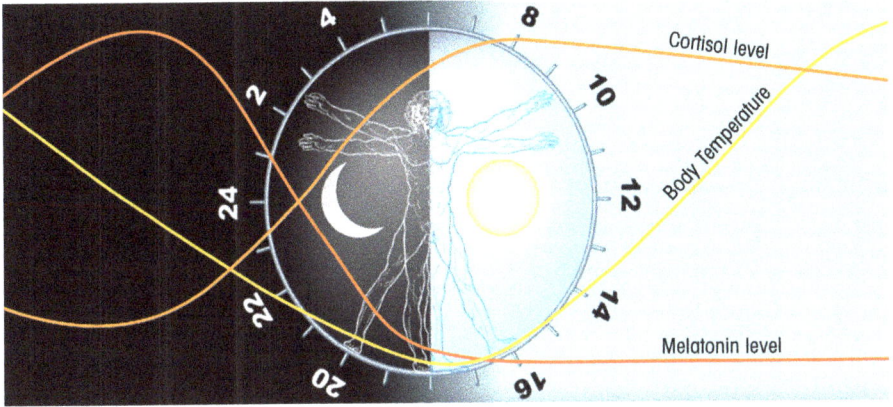

Figure 4. Circadian Rhythm: Day/Night Dynamic and Biological Function

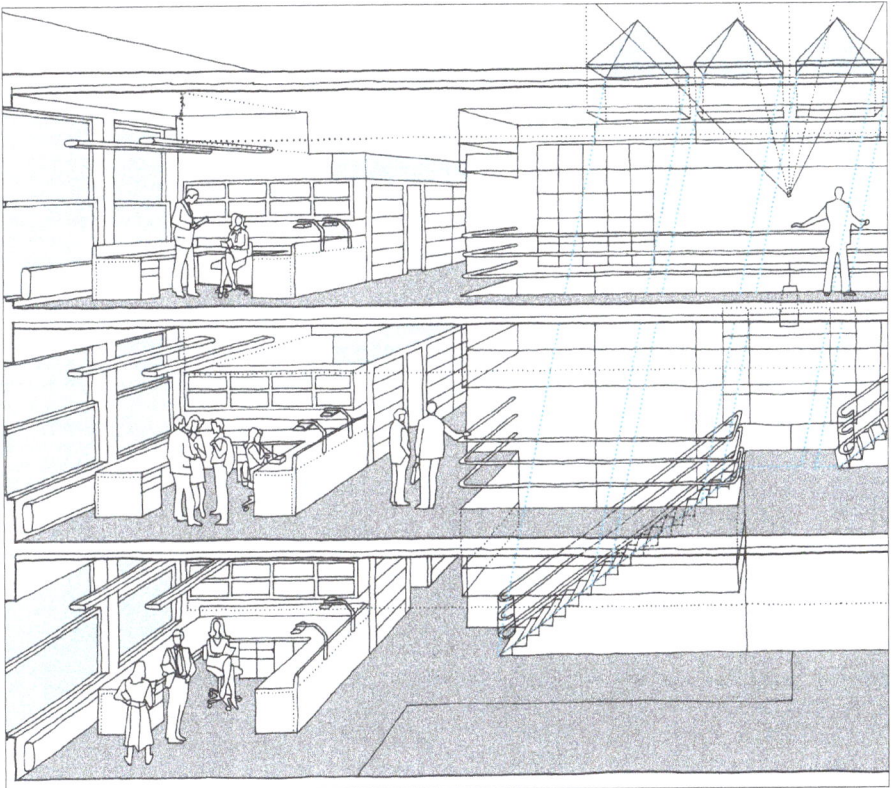

Figure 5. Architectural Integration: Solar Traverse & Work Environment

As a representative of the American Institute of Architects (AIA) and International Union of Architects (UIA) at the UN *Earth Summit* in Rio de Janeiro in 1992, I was able to present these human-centered concepts within the sustainability theme of the Summit. In a follow-up, I hosted an AIA Symposium with the US Department of Energy's National Labs on emerging sustainable innovation. An initiative of one of those labs, Pacific Northwest National Labs (PNNL), led to mapping the role of genes in the Rosetta Stone of Co·GENESIS, the beginning of life as we know it.

Although falling in the domain of bacteria, *cyanobacterium* is commonly called blue-green algae. *Figure 6* shows an abstract pattern of its genes as they are sequentially *expressed* (assigned activities) by RNA during a 24-hour circadian sequence. Jason McDermott, a computational biologist at PNNL, led the team in identifying and mapping the elegant circadian dynamic seen here that uses the sun's energy via photosynthesis to produce sugar during the day (generating oxygen as a waste product) and "fix" nitrogen into usable form at night. Published in *Molecular Biosystems*[3] in 2011, blue-green algae have long been pursued as a sustainable/renewable energy source. Only recently has its functionality been fully recognized as a complex RNA-based regulatory system, enabling the organism to function in tandem with the day-night cycle. Here then, is the single-cell prototype of the human body and the dynamic of our 37 trillion cells over that same 24-hour traverse. As the hours pass, RNA adjusts cellular function in tune with day/night. These ancient single cells, over a period of ±200,000 years, beginning 2.2 to 2.0 billion YBP (years before present), changed the atmosphere of Earth from anaerobic (no oxygen) to the oxygen-laced atmosphere required for complex animals like fish, birds, mammals, and you and me. This lowly single cell is the pivot point to the world we inhabit, still producing 25% of our oxygen.

Figure 7 is an east-west section of Rinker Hall which we designed for the University of Florida in Gainesville (2004). The traverse of the sun's direct beam light animates the central circulation spine, and indirect overhead daylighting reaches the classrooms via the geometry of ceilings, solar screens, and louvers: a design strategy we call "deep daylighting." Maximum connectivity to circadian rhythm is our goal: "A day in the building as a day in nature." The north-south orientation of the building allows occupants to have a sunrise-to-noon or a noon-to-sunset solar traverse with a half-day no-glare direct exterior view. Maximum electrical savings flow from continuous dimming of lighting during daytime.

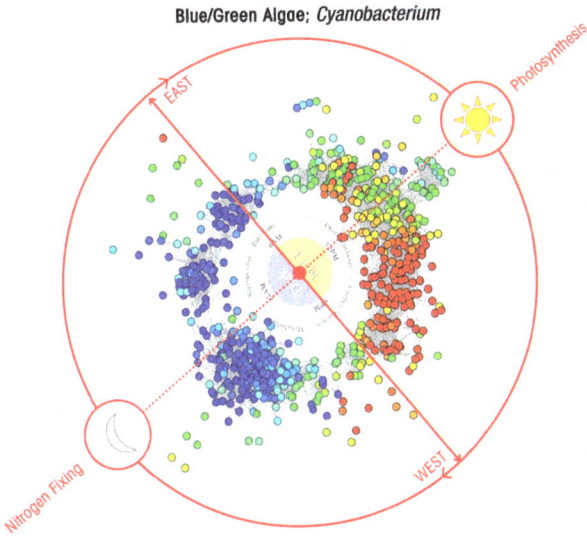

Figure 6. Cyanobacteria Cell: RNA Regulation of Genes Day-to-Night

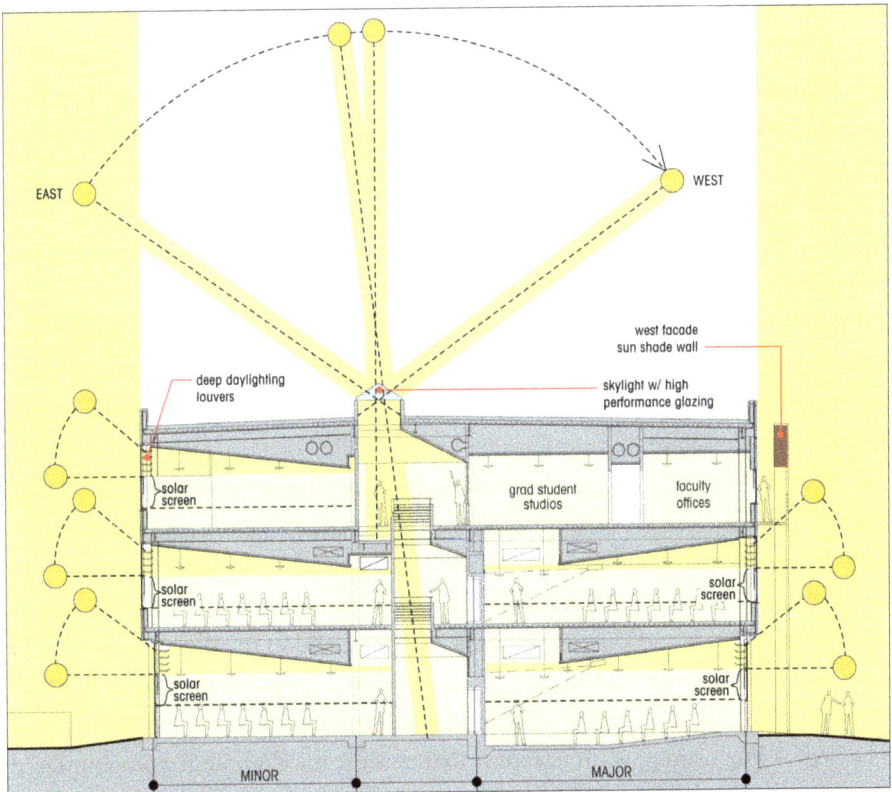

Figure 7. Full Circadian Integration: Direct @ Circulation, Indirect @ Classroom

Because Rinker Hall was new construction, we were free to achieve what we deemed an ideal orientation. But what do we do about existing buildings and, most significantly, what about the massive number of Modernist buildings with their east-west orientations and aversion to solar penetration? What to do with them? Life seldom gives us perfect opportunities to test out such challenges but we got our chance in the 68,000 square-foot Wooster Science Hall at the State University of New York (SUNY) in New Paltz, a prototypical "Brutalist" version of Modernism completed in 1967. Working within the existing structure, we were able to open a 250-foot-long daylight spine the full length of the building, reorganize circulation to get classrooms and labs shifted to the exterior walls, and achieve natural light in all regularly occupied spaces (see *Figure 8*). Even the original building's ideal (in the Modernist vernacular) of north-northeast facing clearstory windows were creating a sky-glare condition (see *8a* and *8b*), which we moderated with louvers at the interior window face and an extended light shelf below. Importantly, this gut renovation, completed in 2018, also incorporates our next-generation architectural responses to human perception-behavior as informed by Co·GENESIS.

NRDC, Rinker Hall, and Wooster Hall produce benefits far beyond increased productivity and the enhanced quality of learning environments we sought. *Figure 9* maps the currently understood chronology of bodily functions attuned to time of day and the on-going balance of bodily systems: thermal balance, hormonal balance, water balance—collectively called *homeostasis*. The effects of "jet lag" when we travel across time zones tell the story: the sudden disruption of our circadian rhythm resulting in our not being able to sleep, eat, or be as attentive as our usual selves. We quickly synchronize to a new locale, but extended disruption has serious health impacts. Modern architectural interiors and social behaviors are creating a physiological imbalance, undercutting our health and well-being.

Seul-A Bae et al, authors of "At the Interface of Lifestyle, Behavior, and Circadian Rhythms: Metabolic Implications" in *Frontiers of Nutrition* (August 2019)[4] are but one among many research teams linking circadian rhythm disruption to metabolic dysfunction: "increasing the risk of obesity, metabolic and cardiovascular disease, diabetes and cancer." In short, when we add in the disruption of REM sleep and memory consolidation, a primary cause of poor health and early death in the modern era is due to the disruption of our metronome: circadian rhythm. The Seul-A Bae team goes on to report that chronic misalignment as encountered by night shift workers, or social behaviors at greatest variance from our biological clock, are associated with the greatest health impacts.

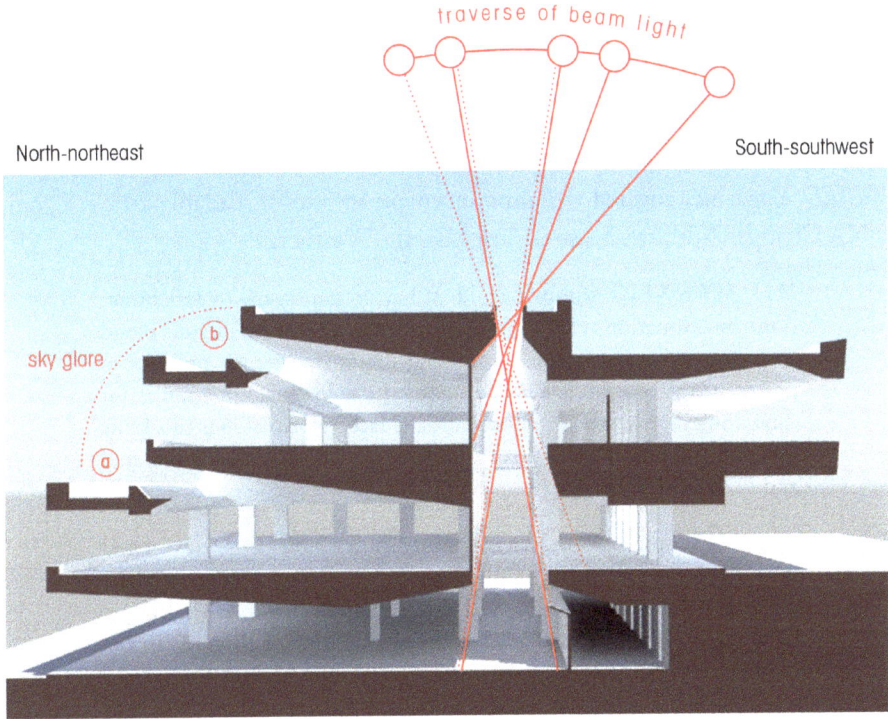

Figure 8. Daylighting Retrofit: Light Slot, Ceiling Geometry & New Exterior

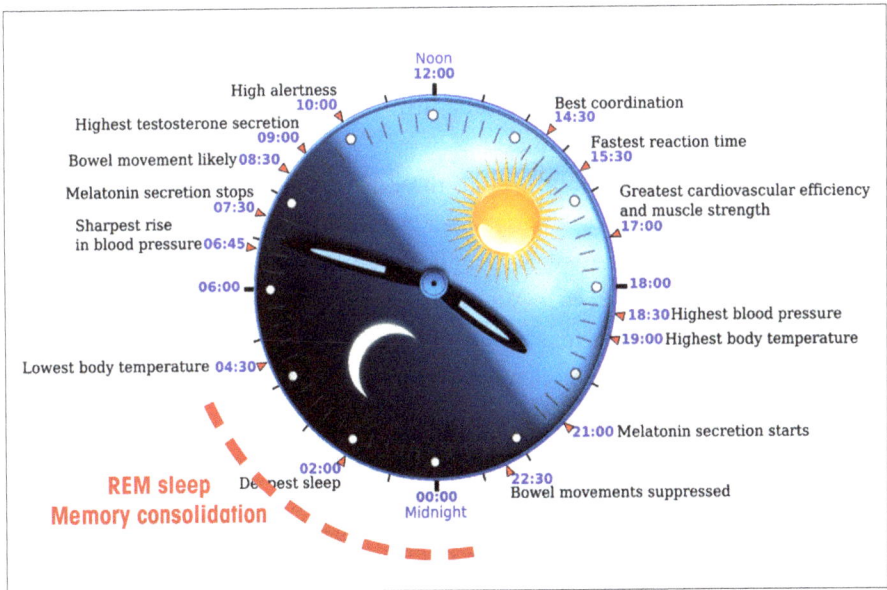

Figure 9. Biological Alignment: Bodily Functions and Time of Day

The role of daily circadian rhythm is well-established, but what about the annual *circannular* rhythm of the seasons? Are humans losing that connection given the isolation of modern life within stable temperature environments far from seasonal change? The answer came on May 12, 2015 in the research journal, *Nature Communications*.[5] On the same day Nick Stockton, a journalist writing for *WIRED* magazine, covered the announcement in a most readable article: "Your DNA Changes With the Seasons, Just Like the Weather:"

> "AH, MY SWEET summer child. What do you know of inflammation? Inflammation is for the winter, when genes uncoil in your blood and messengers send codes containing the blueprints for proteins to protect you from the harsh diseases of the cold. Inflammation is for those long nights, when the sun hides its face, or rain clouds block the sky, and trillions of little T-cells are born to fight the diseases of cold and flu season.
>
> At least that's the news from a new study showing that DNA [expressed by RNA] reacts to the seasons, changing your body's chemistry depending on the time of year. The findings published today in Nature Communications, show that as many as one fifth of all genes in the blood cells undergo seasonal changes in expression (assigned activity)."

While I would change Stockton's DNA attributions to RNA, he does go on to tie the dynamic back to RNA when he notes that under seasonal conditions, genes in the strand are "exposed to messenger RNA (mRNA) which picks up the code and begins the process of making proteins [proteins direct cell function]." For our purposes, the working assumption for the total number of genes in the human body is 40,000. Given the one-fifth estimate, the number of seasonal genes would be approximately 8,000.

Figure 10 is a Section cut through the Entry Hall at Wooster showing two seasonal design elements; on the right is a classic south-facing exterior overhang of the third floor with an east-west line of trees combining to create a cool shaded outdoor "porch" in the summer. When the leaves fall in the winter and the sun drops to its low angle, a thermally-loaded/warmed walking surface assures winter comfort. The most dramatic element is the interior Main Stair and Skylight which are aligned true solar north-south and calibrated so that the trailing edge of beam light passing through the skylight hits the top step at Summer Solstice (see *Figure 11*) and the bottom step at the Equinox (circannular). On every clear day Solar Noon is dramatically marked on the stair by sunlight passing through north/south skylight fins that admit beam light when the sun is near to directly overhead—a brief solar event lasting about an hour, beginning and ending as slivers of light—most brilliant at Solar Noon (circadian).

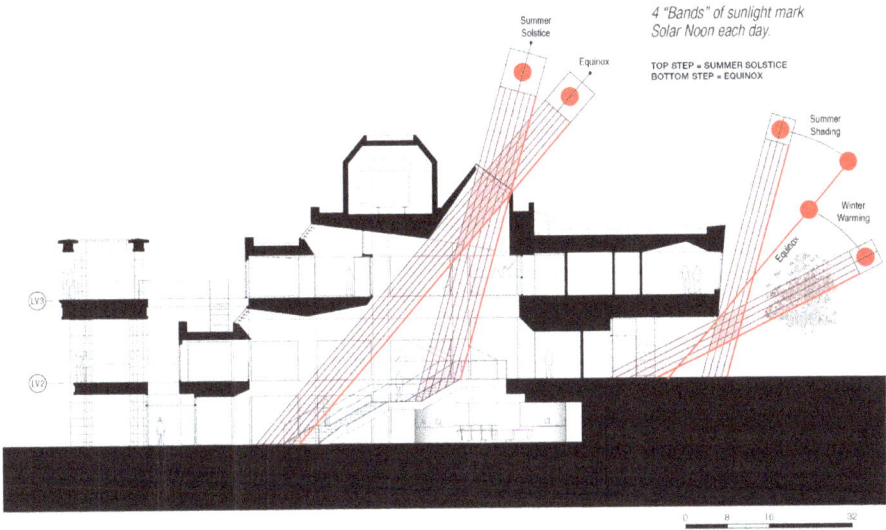

Figure 10. Seasonal Variation of Beam Light: Main Stair Markers and South Porch

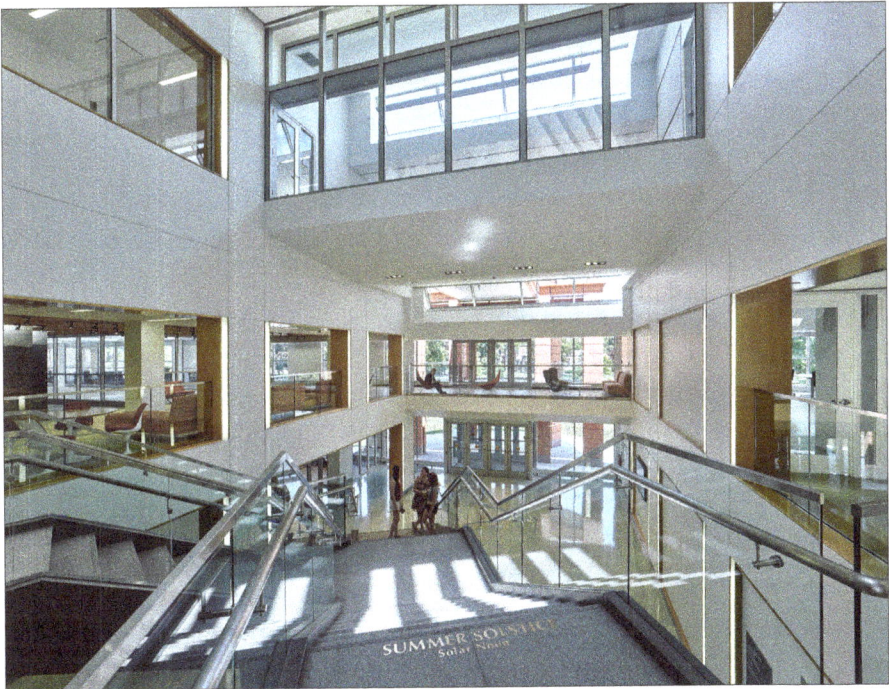

Figure 11. Main Stair at Summer Solstice: 8 Destinations Visible @ Entry to Hall

The architectural design of the 3 story Entry Hall in *Figure 11* is informed by an ancestral survival imperative of humans: seeing and anticipating your enemies (threats) and avoiding being observed or trapped. The concept of *Prospect and Refuge*, put forward by Jay Appleton, the British geographer and academic, in 1975 and formally defined by Annemarie Dosen and Michael Ostwald in the *International Journal of Design in Society* (2013)[6] is consistent with Co·GENESIS. A key is the *overlook* or view into an unfamiliar space (accomplished through a *frame*, a secure edge or spatial sequence imparting a sense of personal control). The photographer in *Figure 11* is on Level 2. On the left and beyond, a balcony edge with seating groups has been provided; Level 3 is similar. Exterior approaches to Levels 1 & 2 provide 1-story transition spaces visible through floor-to-ceiling glass walls. Pedestrians are empowered, deciding whether they wish to enter the building, enter and move along the protected 1-story transition edge, or walk directly out into and through the Entry Hall's 3-story space.

The companion survival imperative is *Way-finding* which, in this case, consists of the eight destinations that are visible upon entry at Level 1, readily apparent to the visitor: Cafeteria, Lounge Seating Overlooks, Student Services Desk, Elevator, Stair to Psychology/Anthropology, Main Stair to Science Quad, the Main Departmental Conference Room, and Connection to Coykendall Science Building. Spatial anxiety, fear of getting lost, and fears of being exposed or stepping into a vulnerable position come down to us from traumatic pre-human episodes of near-starvation, near-death, and Natural Selection (i.e., selective success or failure to reproduce). Ancestral adaptations at our perceptual and behavioral core are the roadmaps for human-centered design.

Figure 12 describes the mid-brain, the RNA MIND uniquely interpreted in each of us. The *hypothalamus*, at the center, is the greatest concentration of RNA male and female imprinted genes in the human body, the coordinator of the neurological and endocrine system, the 'Grand Central Terminal' of perception/behavior. Circadian and circannual timing reside in the SCN/hypothalamus. The right and left amygdala/hypothalamus embodies the FIGHT *maleness* or FLIGHT *femaleness* response. When the Nobel Prize was awarded to Hall, Rosbash, and Young in 2017 for their discovery of the molecular mechanisms of circadian rhythm, we breathed a sigh of relief because our firm had invested over 30 years based on the strong belief, without absolute evidence, that we were shaping architecture to fundamental life forces. Here is where we begin. The RNA brain-within-the-brain, first studied for human-centered architectural design which led to the life sciences, and thirty-four years later, to *The Theory of Co·GENESIS*.

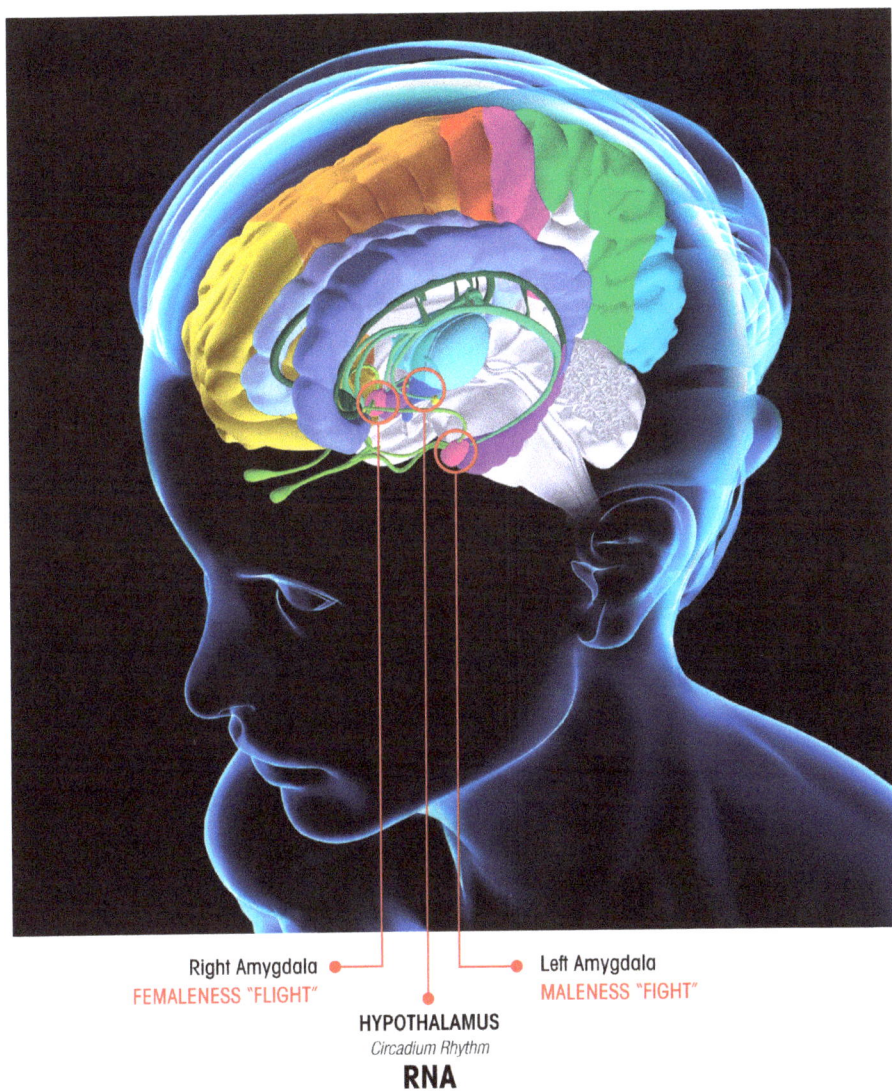

Figure 12. Hypothalamus: Neurological/Endocrine Center of the RNA MIND

CHAPTER TWO

THE BASELINE

RNA (epigenetic) is the determinant force in the creation of our Mind, who we uniquely are. I characterize the Two Minds Theory as an Epigenetic/Genetic Theory of human emergence, which brings us back to an earlier theory—that of the RNA World…that RNA existed as the first formation of life on Earth and that DNA followed.

A Convergence of Two Minds (2015)

In *Two Minds*, I proposed that there are two types of information: RNA (Mind) and DNA (Body), separate pathways into the next generation. Because the first form of life had to evolve a means of replication, it seemed logical that the RNA World Hypothesis was correct, and most scientists in 2015 considered the RNA World a "likely" scenario. However, because I am asserting that Co·GENESIS is a function of RNA primacy, a higher standard of evidence must be met. We need a better answer to the question: "What is the biological origin of life?"

In 2015, some facts in favor of RNA as the first life form were:

1) The main molecular subunit of RNA (ribose sugar) is easier to form than the equivalent subunit of DNA (deoxyribose sugar).

2) In certain viruses, RNA serves as the sole genetic material, while DNA only occurs in the presence of RNA.

3) RNA can form proteins, the body, while DNA cannot.

What had been critically missing was an evidence-based origin story supported in the lab by the replication of a molecular transition from abiotic (non-living) to biotic (living) material; that has changed. During the period 2016-2022 research teams advanced the geochemical and chemical basis for the formation of RNA, including the long sinewy backbone and the attached information bars or nucleotides, as illustrated in the right-hand margin of this paragraph.

RNA

Uwe Meierhenrich, Director of the Nice Institute of Chemistry in France, also directed the team of Cornelia Meinert et al. in publishing "Ribose and related sugars from ultraviolet irradiation of interstellar ice analogs" in *SCIENCE* (April 2016).[1] By recreating the physical and chemical conditions of the primordial universe—a simulation of early Earth's meteoritic bombardment—the Meierhenrich-Meinert team formed *ribose*, RNA's backbone. Starting with frozen meteor and dust compositions, surrounded by water, ammonia, and methanol, the team recreated meteors penetrating the planetary atmosphere. Following the temperature flux, ultraviolet light, and atmospheric pressure, a condensation of organic residues, including ribose, collected on a cold surface.

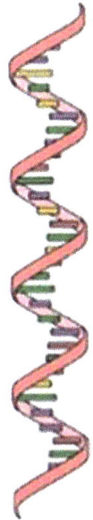

The experiment demonstrated the abiotic (non-living) synthesis of ribose, which is the "central molecular subunit of RNA" (see *Figure 13*). The Meierhenrich-Minert team's research dovetails with "Origin of the RNA World: The fate of nucleobases in warm little ponds" in *Proceedings of the National Academy of Sciences (PNAS)*, October 2017, by the Ben K. D. Pearce team at the Origins Institute in Ontario, Canada, and Max Planck Institute for Astronomy in Heidelberg, Germany[2]. They envisioned a subsequent period of heavy meteoric and volcanic activity, leading to the rise of the continents and the formation of warm ponds of water with seasonal cycles of filling up and drying out (See *Figure 14*). A mathematical model was derived for the evolution of meteor-delivered ribose and organic material during the wet/dry cycles, supporting the emergence of nucleotides, the second building block of RNA molecules. Researchers estimated that RNA, the first form of life, appeared ± 4.17 billion YBP.

Just over a year later, Sidney Becker et al. at the Center for Integrated Protein Science, Munich, Germany, writing in *SCIENCE* (October 2019),[3] advanced the wet/dry cycle concept in an extraordinary demonstration of "chemical pathways that allow formation of life's key building blocks." The researchers were able to show chemical interactions leading to the *concurrent formation of all four RNA nucleotides* as a result of simple rainwater accumulation, wet/dry cycles, sedimentation, and stream flows on the Earth's surface. The Meierhenrich-Meinert derivation of *ribose*, and the Pearce and Becker teams' complementary findings leading to their assertion of the concurrent formation of *nucleotides*, provide our origin story: the rise from abiotic to biotic, via physical, geochemical, and chemical pathways—a foundation for the Theory of Co·GENESIS.

If ancestral RNA came first, how did DNA come to be? In modern times we see the DNA helix, as the fountainhead of human biological knowledge, holding the information for RNA, DNA, Proteins, Perception, the Developmental Stages of Life, Reproduction, Aging, etc. While the helix and its genes are a Code, composed of DNA, there is consensus that ±24% of the genes are *regulatory*, functions of the body carried out by RNA agents which have been transcribed, essentially copied, from the helix by RNA. In the same manner, the 1% *coding* genes are transcribed by RNA to form the body's proteins, and the 1% RNA *imprinted* genes are proposed here as the binary RNA Mind (male legacy/female legacy). In short, if it is true that the operation of the human body is controlled by RNA, why wouldn't RNA (short-lived) have evolved DNA as a long-lived endless source of new RNA and the transgenerational carrier of the encyclopedia of information? Gerald J. Joyce and Biswajit Samanta at the Salk Institute in La Jolla, California, reported that they evolved just such a bridging element in their lab demonstrating the transition from ancestral RNA material to the DNA helix.[4]

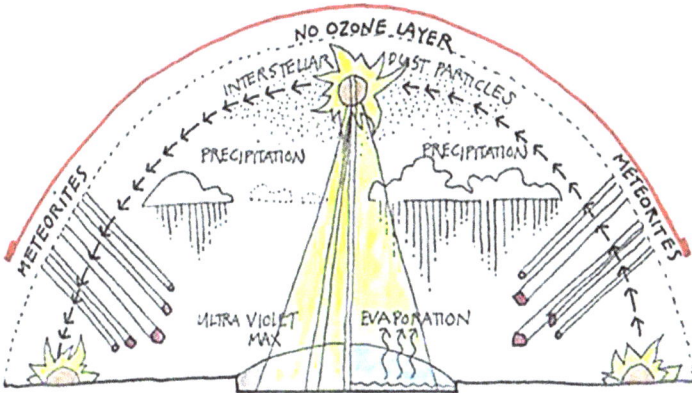

Early Earth: Meteoritic Bombardment 4.54- 4.17 Billion Years Ago

"Our experiments demonstrate plausible physical and chemical environmental conditions that allow for abiotic ribose synthesis."[1]

Meinert et al.,

"Ribose and related sugars from ultraviolet irradiation of interstellar ice analogs" Science (2016) **352** (6282), 208-212. DOI: 10.1126/science.aad8137

Figure 13. Origin of the RNA World (Pre-biotic)

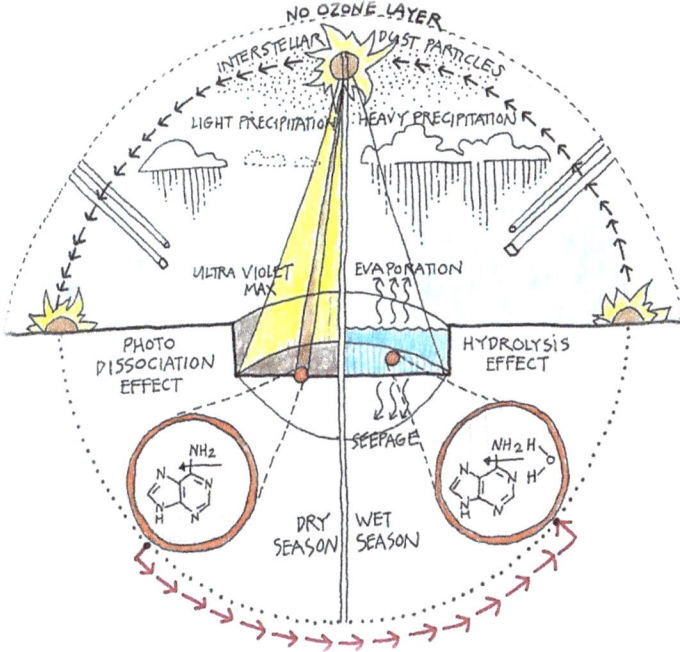

Habitable Earth: 4.17-3.87 Billion Years Ago

"...molecular evolution of nucleobases and subsequent polymerization into RNA [the RNA WORLD]"

K. D. Pearce et al.,

"Origin of the RNA world: The fate of nucleobases in warm little ponds" PNAS (2017); 114 (43) 11327-11332.[2]

Figure 14. Origin of the RNA World (Biotic)

The premise of Gerald J. Joyce, current President of the Salk Institute, is appealing because it embodies the tight yin-yang functional relationship scientists observe between RNA and DNA. *Figure 15* conceptually illustrates the relationship and, although the RNA functional capability was demonstrated by the research team, they point out that the event was in the early life of Earth and is, in that sense, speculative:

> All discussion pertaining to the transition from RNA to DNA genomes is speculative, although *arguably this event is one of the most significant in the history of life.* [Emphasis added] Without the transition to a more stable genetic material, the length of heritable genomes and therefore the complexity of life would have been severely limited…The information content of modern cellular organisms likely would not have been possible without the invention of *reverse transcriptase* [the making of DNA by RNA].

The striking similarity between RNA and DNA in *Figure 15* is obvious: they both have a "backbone" (shown in green and gray-green) with attached rods of information, the "nucleotides" (shown in four colors). The backbones of the molecules are formed from sugars: *ribose sugar* in RNA (the "R" in RNA) and *deoxyribose sugar* in DNA (the "D" in DNA). The chemical composition of the two sugars is so similar that there is just one atom of oxygen difference. The familiar DNA helix, two backbones interwoven, is always the star of the show while RNA is typically shown as we see it here, one strand, appearing to be an irregular half of the DNA helix geometry. This is RNA in its initially transcribed state, when it is being "pulled out" of the helix and before it pirouettes into the many complex forms of active RNA. Our working assumption is that RNA, beginning as the simplest of life forms, transitioned to complex modern life forms via the DNA helix, in which RNA can be multiplied and passed down within a stable molecular structure. As we will see, RNA can quickly call-up thousands of its RNA legions by ordering the cell and its DNA helix to self-copy as needed—quickly meeting the regulatory, repair, and replenishment requirements of the human Body.

Because the DNA helix and the embodied Code of Life are the entire human script, and exist in essentially every cell, a cell can be directed by RNA expression to perform differently in multiple organs and at different stages of life. The genes are the units of heredity, so the two half-sets of your parent's genes, contained in the sperm and egg, were all that was needed to form your set of raw ingredients—the unprocessed first cell, the *zygote*. The 99% RNA genes were mixed male/female for the RNA Body and are recast in each generation. However, the 1% *imprinted* RNA MIND genes (male/female binary), pass through essentially intact, unmixed and immortal, as Plato and other Greek philosophers proposed—the immortal MIND within the temporal Body.

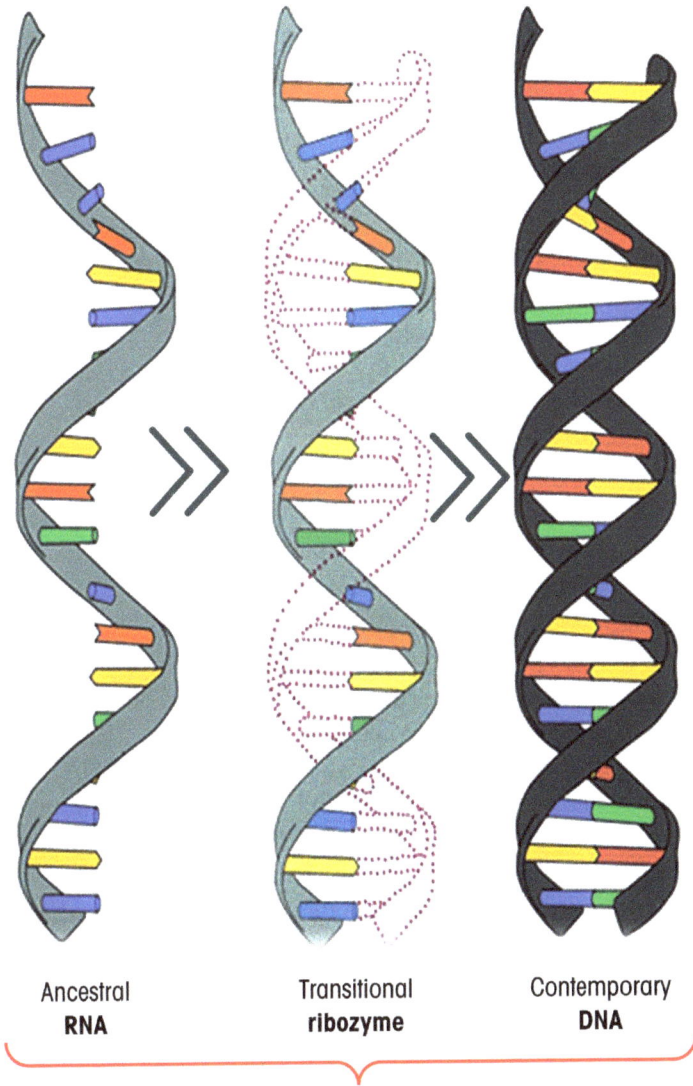

Ancestral
RNA

Transitional
ribozyme

Contemporary
DNA

RNA Synthesis of DNA

Biswajit Samanta and Gerald J. Joyce
"A reverse transcriptase ribozyme"
eLife (2017); **6**: e31153. doi.org/10.7554/eLife.31153[4]

"The present study demonstrates that an RNA enzyme with highly evolved
RNA-dependent RNA activity, can also function as a reverse transcriptase
thus providing a bridge between the ancestral and
contemporary genetic material without the need for proteins."

Figure 15. TRANSITION from the RNA World

Because all information is stored in the helix at the center of each cell, we need go no further to understand the role of RNA. Only RNA can withdraw (transcribe) information from the helix to form protein and the thousands of RNA free agents throughout the body. For our purposes the cell has three basic domains, each housed within the next in the fashion of a nesting Russian doll; RNA works across all three domains.

1) The *nucleolus*, the oldest and smallest of the three domains, resides inside the cell's nucleus; this is the factory where the components of the *Ribosomes* are formed. Ribosomes are the RNA translators of information into proteins. Ribosomes were present in the single cell cyanobacterium before life, as we know it, existed...imagine a barren Earth without mammals, birds, fish, or even plants—before an oxygenated atmosphere.

2) The *nucleus*, globular in shape, is the control center of the cell. Surrounded by a double membrane, RNA enters and exits through its nuclear pores. Aside from the nucleolus, which occupies about 25% of the nucleus, the rest is occupied by the DNA helix in sprawling "spaghetti" form. No information in the helix can be removed, it can only be transcribed (copied) via RNA action.

3) The *cell membrane* is the permeable outer boundary of the cell which encloses the liquid ocean of cytoplasm—a domain of RNA agents forming proteins and regulating bodily function. RNA moves through the surrounding cellular membrane to flow throughout the rest of the body for purposes of extracellular communication. This teeming universe of RNA information achieves the near-simultaneous circadian and circannular alignment and coordination of all bodily functions.

RNA *(RNA polymerase)* is in charge of transcription, the making of RNA molecular agents from information in the helix. The process is called *de novo* because it creates new RNA agents by configuring their nucleotides into an array of complex forms. DNA *(DNA polymerase)* on the other hand, (only when prompted by RNA) simply makes copies of the helix during cell replication. In *Figure 16A*, RNA (polymerase) has formed a small molecule called tRNA from its DNA blueprint in the helix. This active tRNA has two typical differences from its DNA helix twin which has a backbone of *deoxyribose sugar* and is sedated with *thymine* (one of its four nucleotide types). The newly formed tRNA has the more versatile *ribose sugar* for its backbone, and replaces the sedating *thymine* nucleotide with facilitating *uracil*. The tRNA is now liberated to reconfigure and execute its assigned function (see Figure 16B). Even the finished tRNA molecule (see *Figure 16C*) with only 70 to 90 information bars is so complex that it needs a 2-dimensional *Secondary Mapping* in order for us to easily read the connections (see *Figure 16D*).

helix of sugar-phosphate backbone

base pairs

hydrogen bond

RNA

adenine
thymine — uracil
guanine
cytosine

HELIX

bases

(interconnecting information bars)

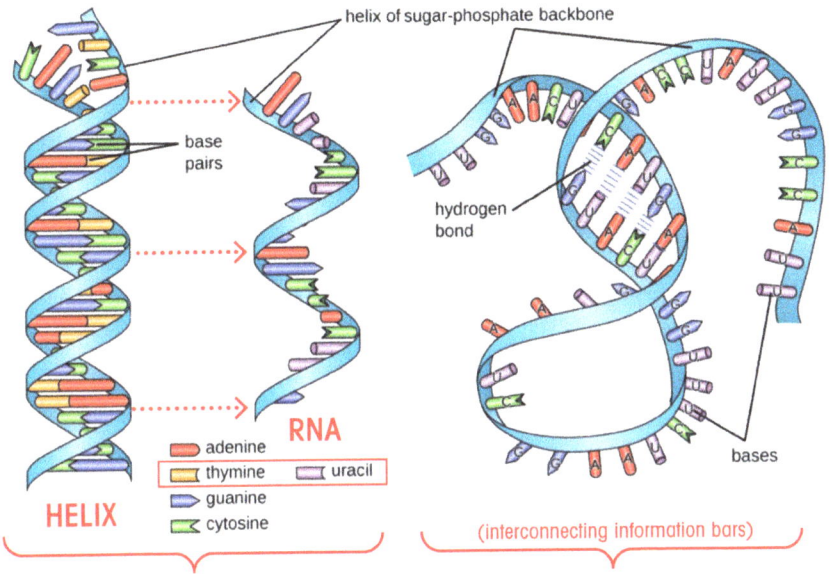

(A) Transcription of the RNA kit

(B) Assembling the RNA molecular kit

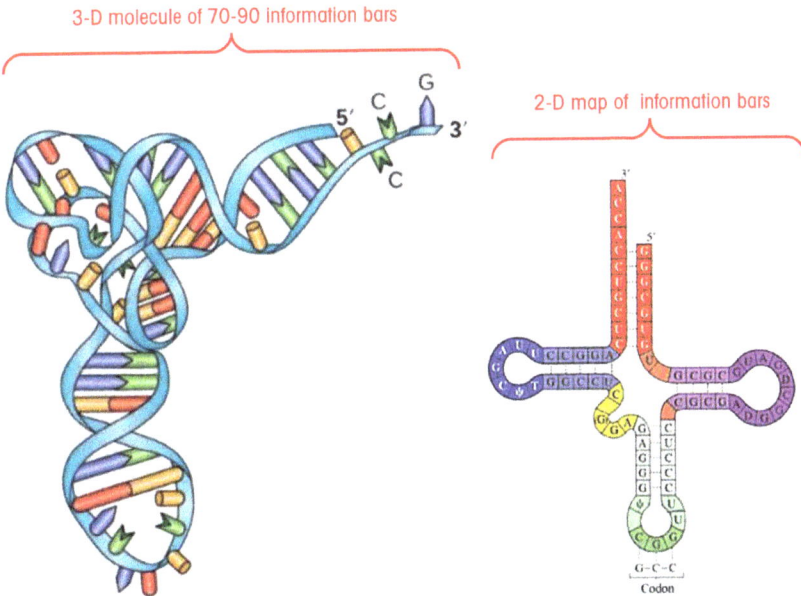

3-D molecule of 70-90 information bars

2-D map of information bars

5′   C   G
3′
C

G–C–C
Codon

(C) Transfer RNA molecule (tRNA) (assembled)

(D) tRNA "Secondary Mapping"

Figure 16. Helix-to-RNA-to-tRNA Functional Molecule

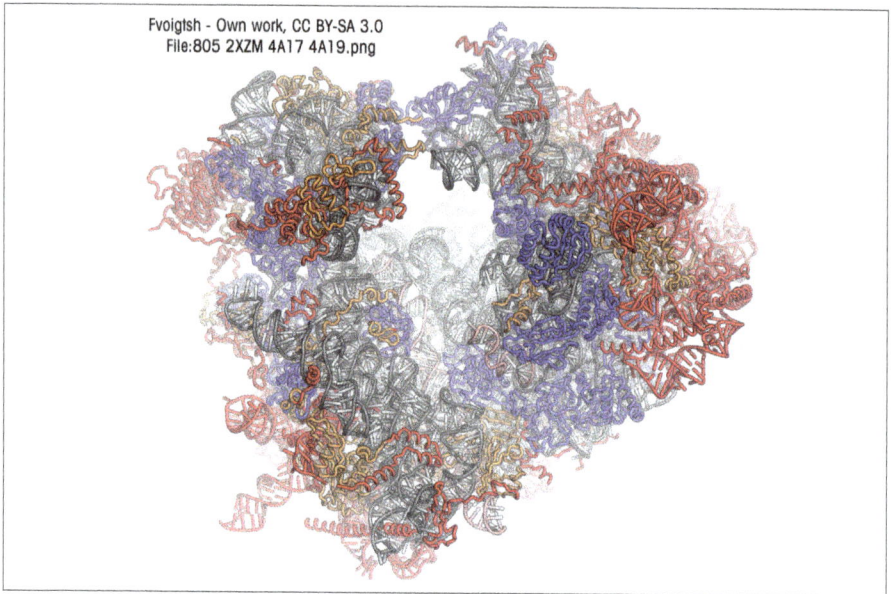

Fvoigtsh - Own work, CC BY-SA 3.0
File:805 2XZM 4A17 4A19.png

Figure 17. Master RNA Organelle - RIBOSOME

Behold in *Figure 17* the stunning 3-dimensional complexity of the Ribosome, the micro-molecular machine that produces protein. There are millions of these RNA "factories" in the cytoplasm of each cell, and they are driven by large and small subunits of yet another RNA molecule: rRNA. The modern version of Ribosome pictured above has evolved proteins (blue range) and expansions of rRNA (red range). The complexity and high functionality of just one of its small subunits, rRNA, (*thermos thermophilus*) is evident in the mapping of *Figure 18.*

Baseline Chapter Two Co·GENESIS assertions are as follows:

1) RNA is the first form of life, consistent with the RNA World Hypothesis as further advanced by the research teams of Uwe Meierhenrich and Cornelia Meinert (2016), Ben K.D. Pearce (2017), and Sidney Becker (2019).

2) DNA was synthesized as a molecular subset by the RNA superset, as advanced by Gerald J. Joyce and Biswajit Samanta (2017).

3) Information exists in two molecular forms: first as the hyper-active RNA molecular agents moving within the cells and throughout the body (extracellular) and, second, as the inactive master DNA coded copy—the helix within the cell nucleus.

4) The DNA helix, effectively a hard drive for RNA, appears to act independently during the replication of helix and cell, however this is a subroutine (as in a computer code) initiated by RNA.

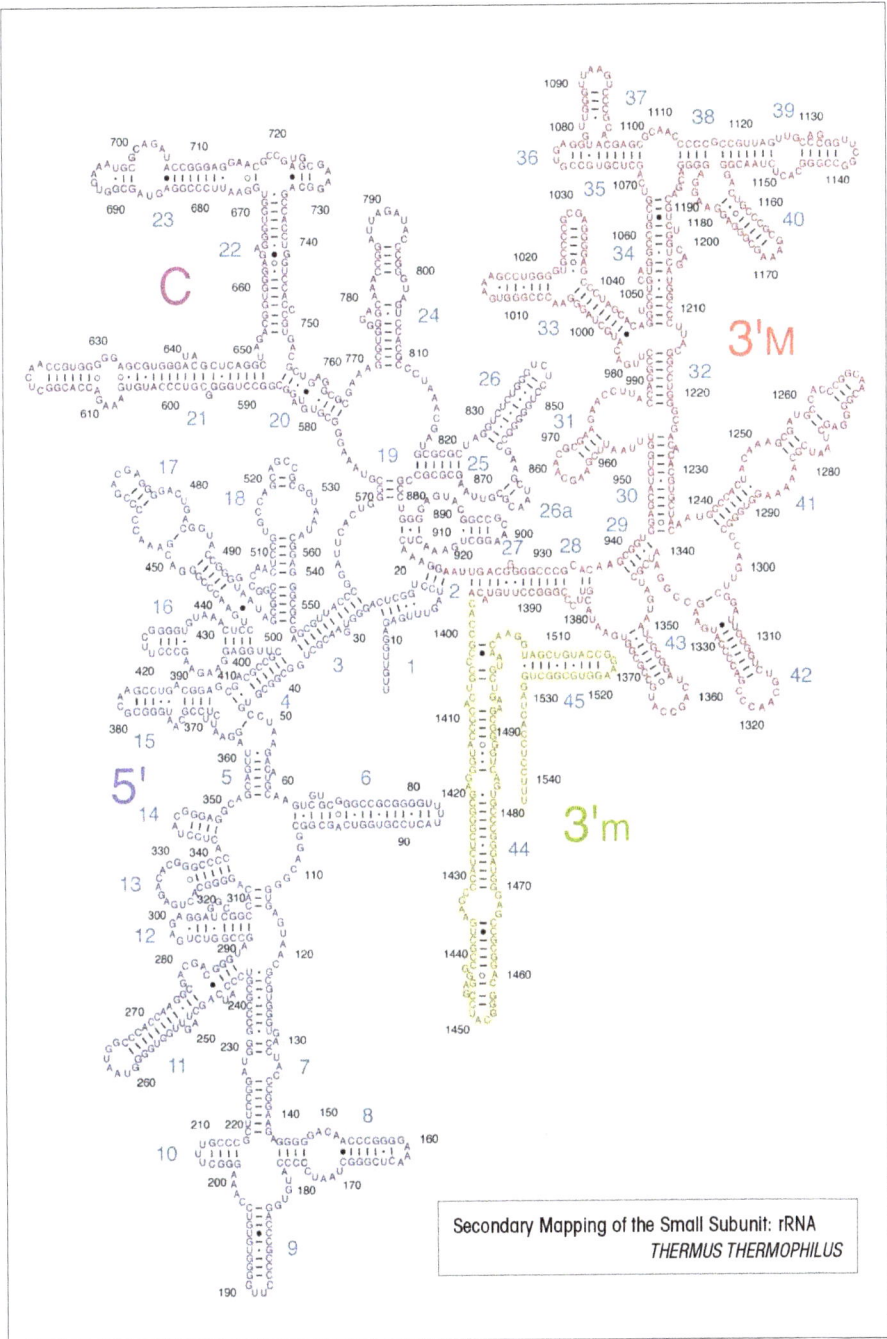

Figure 18. RIBOSOME rRNA Subunit: Over 1,500 Information Bars

THE BINARY

For all the characterizations of DNA's "selfish" genes and their willful actions…genes turn out to be more like birthday ponies that are being epigenetically bridled and saddled by the tags and markers of RNA for their creative pursuits in forming the human mind—a revolutionary reconceptualization should it prove to be correct.

A Convergence of Two Minds (2015)

Co·GENESIS, seven years later, is the anticipated reconceptualization. The helix, containing the Code of Life is often referred to as the Genetic or DNA Code, although it contains all the information necessary for the development, maintenance and reproduction of life so it must, at a minimum, contain the information for RNA, DNA, Proteins and the genetic (forms) and epigenetic (functions) that constitute life. If a code is written in ink, we don't call it an "Ink Code" and so it is with the helix which is far more than just DNA or Genetic. Therefore, I will refer to the helix simply as The Code of Life, or The Code.

Information exists in physical form as organic molecules, RNA and DNA. As we observed, but worth restating here, the information exists in two functional states: *DNA-inactive*, which remains as the helix master copy within the nucleus of our cells, while *RNA-active* builds the Body, controls and operates the cells, and commands the extracellular oceans of the body with legions of RNA molecular agents pulled as needed by RNA from the helix. RNA and DNA can be thought of as reversible states of the same information being driven by the RNA MIND.

Genes are the chapters of heritable information, the parental sourcebooks to be reconciled in offspring. 99% of these genes are mixed to form the two categories of the RNA Body: 1% are coding genes (genetic/body) and 98% are non-coding genes (epigenetic/regulatory agents of body and mind). The RNA Body is a mixture of all ancestors—male and female—and is therefore *unary*. The remaining fraction of 1% of the genes are *RNA-imprinted*, a male lineage and female lineage: two separate rivers of information that interact as a *binary*, our neurological/endocrine system, our RNA MIND. Our parent's half-cells of sperm and egg also contained their life experiences which were passed on as updates to our RNA MIND. If your subsequent male life experiences (via sperm) or your subsequent female life experiences (via eggs) pass to offspring, the updated male legacy will inform the left hemisphere and the updated female legacy will inform the right hemisphere of your children.

The words "mind" and "intelligence," used often in this book, are generally avoided by scientists in favor of *biological mechanisms* and *random chance*. However, in Co·GENESIS we see the evolutionary steps of information-to-knowledge, knowledge-to-intelligence, and finally, two streams of adaptive transgenerational intelligence interacting in real time: *deliberative biological intelligence*. The neuter genes, the RNA Body, undergo dilution and re-expression in each birth, while the Binary RNA MIND passes down essentially untouched, gaining adaptations from the life just lived—two lines of evolving biological intelligence.

How is it possible that scientists have not recognized the primacy of RNA? Why has DNA primacy achieved near-unanimous consensus for over sixty years? The answer is circumstantial evidence. First is the fact that the RNA genes, when in the mode of *stored information*, are effectively invisible within the helix—a DNA molecular composition that is stable and easily studied, Next, the RNA genes and other RNA molecules are active and short-lived outside the helix— much more difficult to study. In addition, the imprinted RNA genes represent only 1% of the genes in the Code and do not follow the the laws of Mendelian Genetics that appear to direct the rest of the Code. Finally, because the Code is written in DNA, one might logically conclude that DNA wrote it. However, a misunderstood event turns this evidence upside-down: *RNA genomic imprinting*. During the formation of the gametes in the fetus, four months in, imprinting splits a portion of the updated binary cells of the Applied RNA MIND, silencing the genes in one half to form the unary sperm or unary eggs—re-establishing the separate evolutionary pathways of the RNA MIND.

We humans are *placental mammals*, having emerged 65 million years ago after the extinction of the dinosaurs at the end of the Cretaceous Period, however, it only takes the six most-recent generations (G1-G6) illustrated in *Figures 19 & 20*, to understand the two information flows leading to us. We begin with *Figure 19* which reveals the simplistic nature of the 99% RNA Body. Your father's legacy is on the left, your mother's is on the right, and each parent brings 31 ancestors including themselves to your mixed genes. 62 male and female gene sets are mixed to form the genes of your *diploid* cells—neuter (neither male nor female). In *Figure 20* of the same six generations we see that 52 ancestors have been excluded in a process called "silencing" by RNA (genomic imprinting excludes the hollow circles). Five direct-descent males in your father's line and five direct-descent females in your mother's line inform the respective halves of the binary RNA MIND.

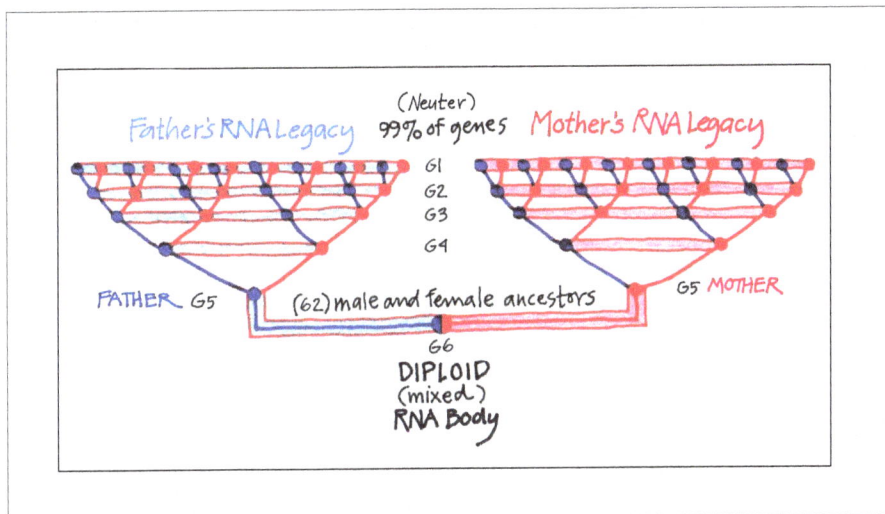

Figure 19. 99% of GENES: Mixture of all Ancestral Genes

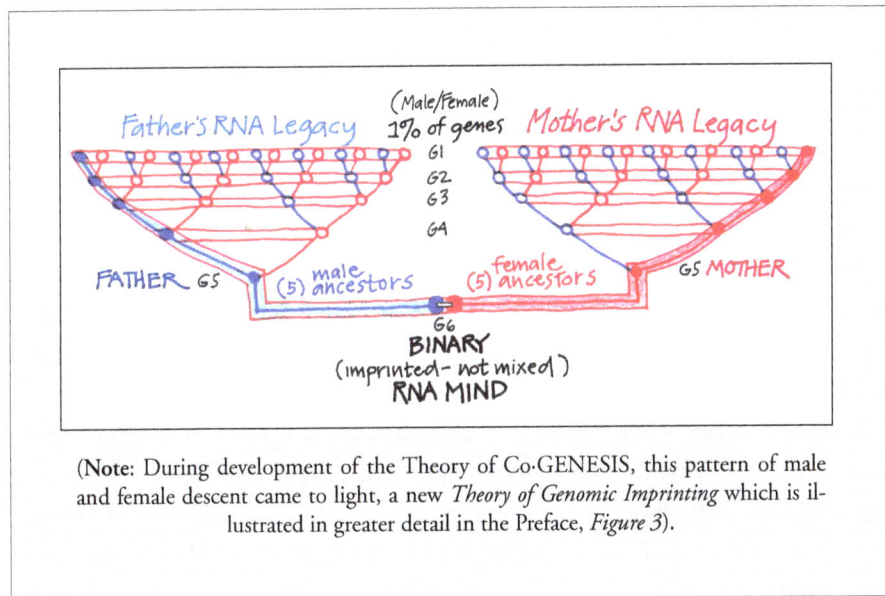

(Note: During development of the Theory of Co·GENESIS, this pattern of male and female descent came to light, a new *Theory of Genomic Imprinting* which is illustrated in greater detail in the Preface, *Figure 3*).

Figure 20. 1% of GENES: Male Only and Female Only Ancestral Genes

Although our mammalian line goes back ±65 million years, RNA-imprinted genes go back much further. Imprinted genes occurred before the last common ancestor of plants and animals, and therefore can be found in contemporary flowering plants as well as wide-ranging animal species, from microscopic mites to whales. Therefore Co·GENESIS, the randomized distribution of perception-behavior and adaptive characteristics through separate male and female lines of descent, may well be inherent to all sexually-reproducing life forms; a subject we will explore further in Chapter 7: Significance and Meaning.

There have been many Theories of Genomic Imprinting over the last thirty years. In their review in *Heredity* (April 2014) M. M. Patten et al. identify "the three theories that have best withstood theoretical and empirical scrutiny"[1] as listed at the top of *Figure 21*. The most widely accepted is the Kinship Theory which attributes the male or female imprinting of the genes to the competing interests of the parents: the parental conflict hypothesis which suggests that the father seeks maximum growth for the fitness of his offspring (even at the expense of the mother) while, to the contrary, the mother seeks to conserve resources for her and future offspring by limiting growth. The chart in *Figure 21* is taken from "Selfish Genetic Elements" by J. Arvid Ågren and Andrew G. Clark in *PLoS Genetics* (November 2018).[2] Utilizing often-cited support for the Kinship Theory, the oppositional roles of the *insulin-like growth factor 2* imprinted genes—*Igf2* male-expressed (increased growth) and *Igf2r* female-expressed (reduced growth)—the effect is illustrated in mice. The authors point out that imprinting is a violation of Mendelian Genetics and "seems like a *maladaptive phenomenon*." These "selfish" genes are seen as rogue elements, acting to assure their own transmission generation-to-generation, without evolutionary benefit. Richard Dawkins popularized this concept in the 1970s.

Importantly, the M. M. Patten team observes that none of the three theories has developed the evidence necessary to achieve predominance. The Patten team's critiques "highlight ambiguities in and overlap between the predictions they make, with a goal of motivating further research." The Patten team goes on to encourage "tests to discriminate between these alternative theories for why particular genes are imprinted." In short, genomic imprinting has been long-recognized but is, to this day, not understood.

A great place for us to begin is with the same *Igf2-Igf2r* binary gene pairing. In a reversal of the theories above, Co·GENESIS asserts that, far from being rogue elements in conflict, these genes are anchor points in a regulatory range of scale from large-to-small. In this instance, the RNA binary is sizing the individual body.

GENETIC IMPRINTING THEORIES

I. Kinship Theory/ Parental Antagonism
II. Sexual Antagonism Theory
III. Maternal-Offspring Co-adaptation Theory

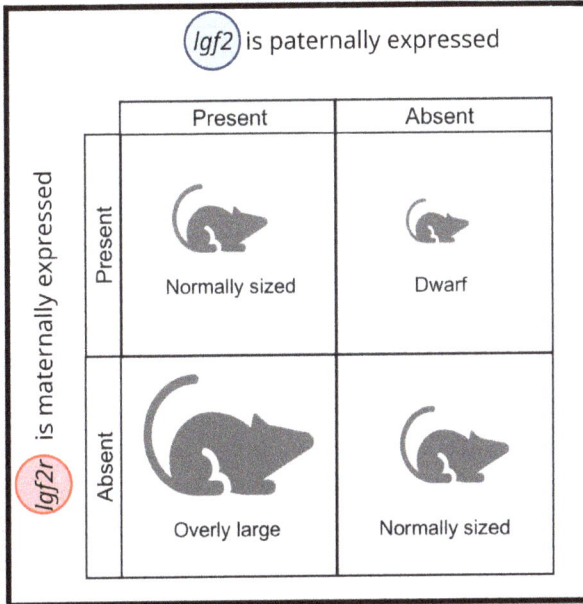

"Selfish Genetic Elements"

Agren and Clark (2018) "Selfish genetic elements" *PLoS Gen* **14** (11): e1007700. doi.org/10.1371/journal.pgen.1007700[2]

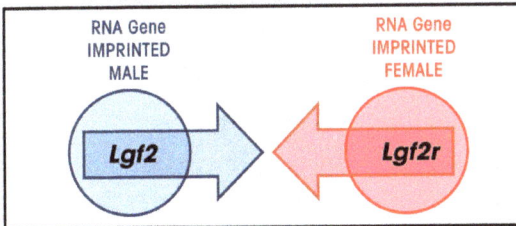

"Another sort of conflict that genomes (genes) face is that between the **mother and father competing for control of gene expression in the offspring**...conflict between the two genomes might be driving the evolution of imprinting."

Figure 21. Theories of GENOMIC IMPRINTING

See *Figure 22A* which recasts the "conflict" in mouse body size as a range for adaptation to the environment. For example, on the Indonesian Island of Flores, *Homo Floresiensis*, a primate closely related to humans (just over three feet tall) and an extinct species of dwarf elephant are thought to have evolved their body size to survive the low resource/low calorie environment. Tucci et al., writing in *SCIENCE* (August 2018)[3] concluded that their spare diet was the natural selection driver, identifying an RNA-imprinted pair of genes as the adaptive factor: "current evidence points to a critical role of FADS (FADS1-FADS2) acting as an evolutionary 'toggle switch' in response to changing diet."

Homeostasis is the maintenance of an a steady internal state of the body's physical and chemical processes, essential for life. One of those processes is the maintenance of thermal balance illustrated in *Figure 22B*. Body temperature must be regulated around 98.6 degrees Fahrenheit, referred to as the set point. As heat is gained or lost due to environmental conditions, the multiple RNA genetic "toggle switches" act to add or subtract heat through various biological strategies (opening or closing pores, sweating, shivering, etc.) to correct to the set point.

Two candidates exist for authorship of the Code of Life: the *Unary* subset of the RNA Body which goes through dilution, mutations, and re-expression of functions from generation-to-generation, or the *Binary RNA MIND,* a two-factor system like a computer code, and steady as a rock generation-to-generation. Right out of the box we have foundational evidence for the primacy of RNA. See *Figure 22C-Binary Code.*

The location and concentrations of RNA imprinted genes in the human body point to an RNA hierarchical network. Imprinted genes occur in clusters of 2 to 20 genes called *imprinted control regions* (ICRs) which are interconnected across the far reaches of the body. The number of imprinted genes is low in the organs and body tissues, but reaches the greatest density in proximity to the brain. The midbrain and *hypothalamus,* the center of emotional and cognitive behavior, have the greatest concentration. The RNA network operates from the brain and executive ICRs down to the thousands of RNA agents (molecules) that stitch this universe together. Teeming throughout the body's liquid pathways, and working in concert with hormonal (endocrine) and electrical (neurological) signaling, these tiny submersibles are constantly moving in and out of the large cellular boxcars, the manufacturing plants for protein, to achieve the near-instantaneous nature of our functions. Driving this network toward survival and reproduction are the timely and often insistent RNA signals of *hunger, thirst, sleepiness, sexual attraction, etc.*

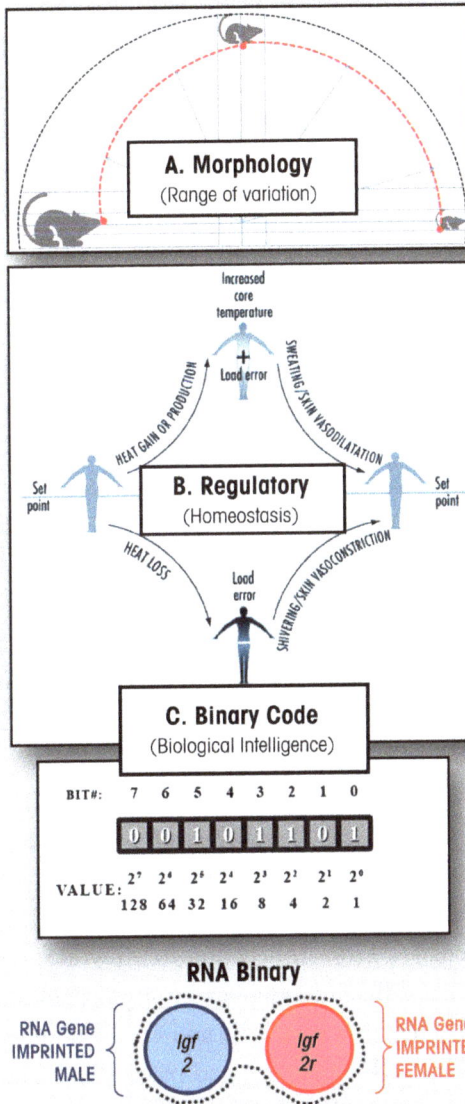

A. Morphology
(Range of variation)

Increased core temperature

B. Regulatory
(Homeostasis)

HEAT GAIN OR PRODUCTION

SWEATING/SKIN VASODILATATION

+ Load error

Set point

Set point

HEAT LOSS

SHIVERING/SKIN VASOCONSTRICTION

Load error

C. Binary Code
(Biological Intelligence)

BIT#:	7	6	5	4	3	2	1	0
	0	0	1	0	1	1	0	1

VALUE:	2^7	2^4	2^5	2^4	2^3	2^2	2^1	2^0
	128	64	32	16	8	4	2	1

RNA Binary

RNA Gene IMPRINTED MALE — *Igf 2*

Igf 2r — RNA Gene IMPRINTED FEMALE

The Theory of Co·Genesis

Recognition of RNA Imprinted genes (male/female) as the
binary dynamic of human life function.

Figure 22. GENOMIC IMPRINTING as a Binary Code

The physical order flowing from genomic imprinting and the separate legacies of male and female life experience, is illustrated in *Figure 23*: MALE and *Figure 24*: FEMALE. The merging, or more accurately reconciliation, of the two halves of the 1% RNA parental genes inform the Neurological-Endocrine System (blue and red) of the body, the *applied RNA MIND* unique to each individual. The gametes constitute the *anticipatory RNA MIND*—the 1% in male (sperm) or 1% female (egg) that will record and can carry forward life experiences and inform the next generation. In contrast, the neuter RNA Body (the outer shell) has no such "memory."

Why are the pathways for Mind and Body so different? This is a central question that we will answer over the course of this book. The short version is that the RNA Body evolves toward the best fit within the surrounding physical environment to best compete for resources and reproduce. By contrast, the RNA MIND, the center of perception and behavior, has an indeterminant challenge: having to perform in unknowable future environments which can change abruptly from peace to war, or to natural catastrophe and famine. A resilient species must possess the *maximum variation of individual minds* to survive and thrive in future extremes, always including those who can best compete in threat or opportunity. Changes in circumstance can occur instantly and more than once in a lifetime so there is no time for a species to evolve the Minds needed to survive—they must be continuously present.

Figures 23 & 24 are a master course in the conservation of energy. One would think that the Code of Life contains a full set of information for a male and a female body, not to mention the 7.8 billion unique minds that exist in the 2023 world population. However, the Code has a single universal (neuter) body for males and females. From conception to ±6 weeks, the human embryo is formed without sexual differentiation—an enormous amount of information turns out to be shared. The Code need only address the unique sexual expression of the body, primarily through hormones, and place it in the range of body sizes. This explains the mystery of male nipples (universal body), and the viability of transgender reassignment. But the greatest model for the conservation of energy is the bi-hemispheric brain: two separately evolving minds connected in each birth. Through the variation of wiring (interconnectivity) a minimal expenditure of metabolic energy is made to achieve the near-infinite ranges of perception/behavior. In this reassembly, our uniqueness is achieved: the spectrum of sexual attraction, maleness/femaleness expression, and talents: we are each, one of a kind. To understand the different natures of the hemispheres, we turn to the pre-human primate mind, the sourcebook for the human primates that followed.

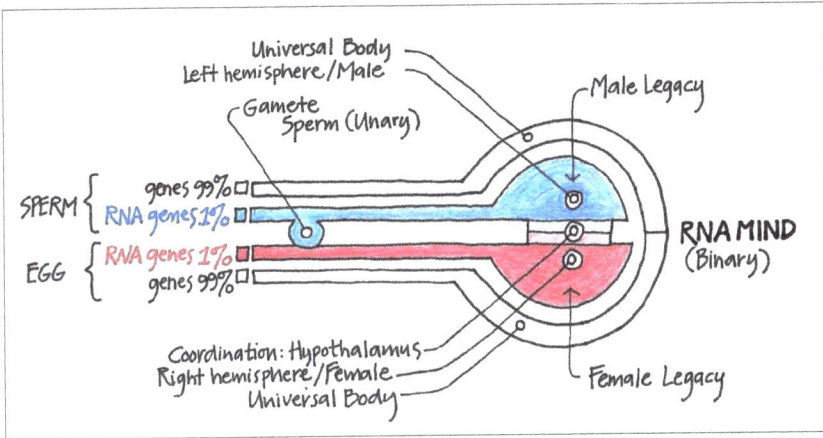

Figure 23. The Male Binary RNA MIND

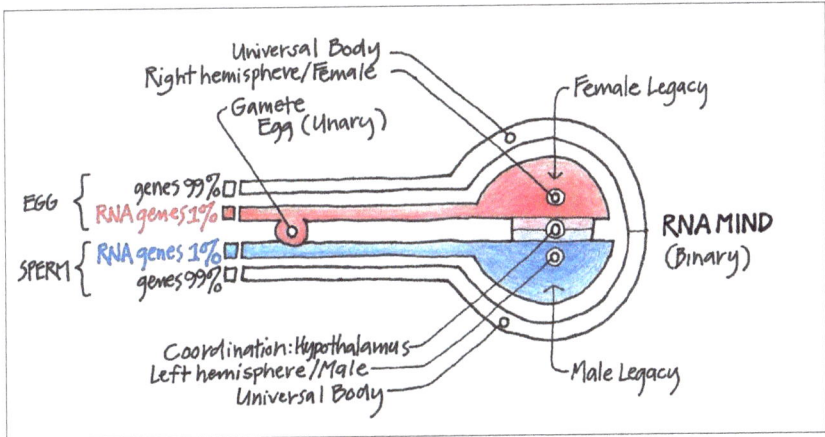

Figure 24. The Female Binary RNA MIND

The earliest evidence in the fossil record for the Primate Order dates back to the *Paleocene Epoch* about 55 million YBP. Our direct pre-Human primate line, the genus *Homo*, arrived 2.8 million YBP in Africa during the *Paleolithic Period*, the age of hunter-gatherers who used fire and hand-chipped sharpened stones for their tools and weapons. *Homo sapiens*, also called anatomically modern humans (AMH), emerged within the genus *Homo* ± 300,000 YBP. However, ±100,000 YBP, a singular evolutionary transition, evident in the archeological and biological record, changed the order of life on Earth. The underappreciated significance of this event, the evolution from the instinctive-subconscious Mind to the overlay of deliberative-consciousness is further developed in Chapter Four: Humankind. For our purposes here, we need only take note of the Primate Order, and consider their ancestral male and female life experiences: two legacies that have formed and informed the Human Mind.

Figure 25 illustrates a key assertion, the pre-human sourcebooks of the Human Mind. Life experiences of pre-humans come down to us organized as left (maleness) and right (femaleness) hemispheres. This does not suggest that males or females always behaved in a certain way, quite the opposite, because Co·GENESIS confirms that both these perception/behaviors are expressed in men and in women. Humans are the only primates to have gained conscious access to these hemispheres.

Pre-humans had their individual mindsets, their various ways of seeing the world (instinctual and subconscious) set for life. Their competitive and often violent environment limited them to the protective structure of small familial bands comprised of ±5-15 individuals bound by blood or sexual relationship—neurobiological bonding via the power of oxytocin and vasopressin (pair and parental bonding). Biological willingness to fight to the death to protect the band was inborn. This hard calculus created two very different life experiences for pre-human males and females. See *Figure 25*.

Evolution of the LEFT Hemisphere: Males in combat who expressed empathy for the enemy, assisted wounded enemies, hesitated to act if uncertain of the identity of an enemy, or befriended an enemy, tended to be killed. Left hemisphere: *Decisive, Aggressive, Act First, No Remorse, No Self-Doubt, Today-focused, Low Emotional Intelligence, and Defensive of Status Quo.*

Evolution of the RIGHT Hemisphere: Females seeking acceptance in new bands (incest avoidance) had to diffuse hostility and establish bonds with others. Under assault or rape, those who fought back or physically resisted tended to be killed. Right hemisphere: *Empathetic, Cautious, High Emotional Intelligence, Future-directed, Socially adept, Flexible, Creative and Cunning.*

55 MILLION YEARS AGO: THE PRIMATE ORDER

All primates have a Bi-hemispheric Brain. Each Hemisphere is
informed by ancestral experience: Male/left, Female/right

(PRE-HUMANS = Non-HUMAN PRIMATES)

PRE-HUMAN MALE
Defender to the death

1) Protect perimeter of the band for
the extended rearing of children
(Community).

2) Raid and/or kill others to
provide needed land and
provisions for the band.

3) Remain with band for life
(status quo).

Life Experience
Bias to NOT TRUST, No Empathy
decisive in taking action, sees black
and white no gray areas, bold,
aggressive, conservative, protect the
status quo, analytical, ego-centric,
low emotional intelligence, impatient.

PRE-HUMAN FEMALE
Survivability of lineage

1) Migrate to other bands
(Avoid incest).

2) Reconcile opposition to gain
acceptance in the new band.

3) Survive kidnapping,
assault, and rape for
long term goals
(Future-directed).

Life Experience
Bias to TRUST, Empathy
emotional intelligence,
holistic thinking, cautious,
sympathetic, liberal,
innovative/creative, adaptable to
change, sense-of-humor, social,
and guile.

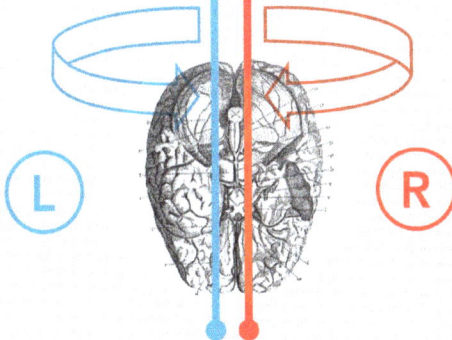

(L) (R)

Figure 25. PRE-HUMAN ADAPTATION of Perception-Behavior

Genomic imprinting is key to maintaining the binary mind across generations and although it was thought to be unique to humans and mammals, it has recently been found in essentially all sexually reproducing organisms. Left-right asymmetry and imprinted genes in the nervous system occur in round worms, fruitflies, and mosquitos, all the way up to pre-humans—variation in perception of the world at the instinctive-subconscious level. We humans, uniquely, have the luxury of consciously examining our predispositions (Nature) and can strive to adjust our behavior (Nurture). However, at the perception-behavioral extremes of minimum empathy (psychopaths and sociopaths) or maximum empathy (saints and heroes) there is almost no flexibility. Predispositions of the instinctive-subconscious Mind, such as sexual attraction are, with few exceptions, dimensions of personality for life. Yes, the instinctive-subconscious is still here today, interacting with the deliberative-conscious overlay.

Iann McGilchrist, literary scholar and psychiatrist, synthesized the neuroscientific evidence related to the hemispheres and rendered a brilliant interpretation in *The MASTER and his EMISSARY: The Divided Brain and the Making of the Modern World* (2009).[4] McGilchrist offered a new perspective with wide-ranging insights on the science of the divided brain: two worldviews in competition for power. In writing *Two Minds* (2015), I interpreted McGilchrist's characterizations within six divisions and chose to order them in an oppositional manner...the only way I could get my mind around the concept. Shown as *Figure 26*, one can imagine each of these columns as the theoretically complete 1% RNA MIND in its two parts as they exist in baby sperm and egg before, many years later, the roulette wheel is spun and there is a distribution of powers, dynamic interconnections, and emphasis that form the next generation. No one could possibly process all the oppositional information in this 1.0 billion year sourcebook; of necessity the individual is an edited and reconciled version drawn from these two massive sourcebooks of the ancestral RNA MIND resulting in a unique variation.

The consistency between the pre-human primate behaviors in *Figure 25*, and the two *ways of being in the world* described by McGilchrist in *Figure 26*, is stunning. I used these parallels in *Two Minds* and repeat them here because I can't find more compelling examples. The only anomaly in neurological distribution occurs under SKILLS in the femaleness right column; 3-D visualization is a predominately maleness modern skill (engineering, and parallel parking). This exception, I speculate, may be a stranding in the right hemisphere due to rapid changes and expansion of brain size in Primates and the "sudden" male reassignment to protect a massive nest, eventually measured in square miles.

MALENESS "Emissary" **Hemisphere L**	FEMALENESS "Master" **R Hemisphere**
WORLD VIEW	**WORLD VIEW**
OBJECTS, NUMBERS, WORDS (Reductionist)	Holistic, the WORLD in CONTEXT (Inclusiveness)
INITIAL STANCE	**INITIAL STANCE**
Frontal lobe BETRAYAL EXPLOITATION	Frontal lobe TRUST EMPATHY
CHARACTERISTICS	**CHARACTERISTICS**
AGGRESSION ANGRY TERRITORIALITY COMPETITION, RIVARY MECHANISTIC INFLEXIBILITY to read emotion of others INABILITY EXPLOITATION UNIFORMITY See another point of view UNWILLING UNDUE Cheerfulness STATUS QUO INTOLERANCE PROJECTION OF POWER	HUMOR SOCIAL ALTRUISM COOPERATION HUMANESS FLEXIBILITY ABILITY to read the emotions of others BONDING (Genuine) UNIQUENESS WILLING to see another's point of view UNDUE Sadness/Depression INNOVATION SENSE OF JUSTICE/Righteous Indignation SPIRITUALITY
FEARS	**FEARS**
ANXIOUS APPREHENSION Fear of uncertainty Fear of lack of control	ANXIOUS ARROUSAL Reading a sad story = sorrow Melancholy–sadness response
SKILLS	**SKILLS**
TECHNOLOGY Performance/MUSIC Linear MATHEMATICAL REASONING Execution/Manipulation of NUMBERS LANGUAGE/VOCABULARY	SPACIAL Interpretation (3-D) MUSIC/Innovation and creation ARTISTIC/VISUAL COMPOSITIONAL/Insights CREATIVE WRITING and VERBAL INTONATION
INTERESTS	**INTERESTS**
ACQUISITION OF MATERIAL MEANS ORGANIZATION/BUREAUCRACY PLANS, MAPS, STRATEGIES INFORMATION and Gathering	ATTENTION to: SEX, ATTRACTIVENESS and AGE FOOD/SOCIAL INTERACTION LITERATURE MUSIC and the Arts

Adapted from Iain McGilchrist, *The Master and His Emissary (2009)*[4]

Figure 26. NEUROLOGICAL PREDISPOSITION of the Hemispheres

Understanding the reconciliation of these oppositional minds is a matter of reverse engineering from the range of personalities we encounter in life. Mindsets that reoccur as a consistent proportion of the population (not just the ones our society calls "normal") are keys to understanding the full evolutionary dynamic. One billion years of sexual reproduction in our line of descent has granted our species absolute dominance, so the human Code of Life and the repetitive pattern (social order) of mindsets we observe are likely to represent the best outcome within the evolutionary dynamic we seek to understand. Although the exact percentages are a subject of debate, an approximate distribution of mindsets that repeat are as follows: Genius (0.25%), Gifted (2.2%: IQ 130+), Savant (1%), Bipolar (2.6%), Autistic Spectrum (1.5%), Psychopath/Sociopath (4%), Heterosexual (96%), LGBTQ+ (4%). Another stable ratio of mindsets is between those seeing a left-hemisphere world of threat (Conservatives 40%) or a right-hemisphere world of opportunity (Liberals 40%). In the middle, between left and right hemisphere influence, are the persuadable mindsets that round-out the repetitive proportions of the political context (Independents 20%).

These crude divisions point toward a universal neurological pattern but miss the subtle nature of our mindsets. A relevant description that has stuck in my mind for over twenty years is that by Paul Ehrlich in *Human Natures* (2000):[5]

> "…genes, environment and gene-environment interactions all play roles in generating gender differences…a range of genders in human beings, not a series of sharp boundaries."

Figure 27 is my speculative "range of sexual attraction" set by a simple RNA MIND algorithm, acting across all male and female mindsets which are shown as circles. Each mindset has a different ratio of maleness (**blue**) to femaleness (**red**). At the top in red is the Maximum Femaleness Influence of the Right Hemisphere and at the bottom in blue is the Maximum Maleness Influence of the Left Hemisphere. Females are on the right and vary from Max Femaleness (top) to Min Femaleness (bottom) and males are on the left and vary from Max Maleness (bottom) to Min Maleness (top). F MAX (top) is the *Alpha Female* and M MAX (bottom) is *Alpha Male*. Moving clockwise, the maleness/femaleness traits begin to balance, reaching 50/50 at mid-point. The neurological gender (NG) of mind *opposite to anatomical gender* then increases to *Omega*: where the opposite NG dominates and gender change of the body (transsexuality) may be sought. LGBTQ+ (non-binary) appears to be expressed in the last 4% of each side, but because it is a range, there is actually a continuous variation of sexual attraction across all heterosexual males and females that eventually expresses as fully non-binary: LGBTQ+.

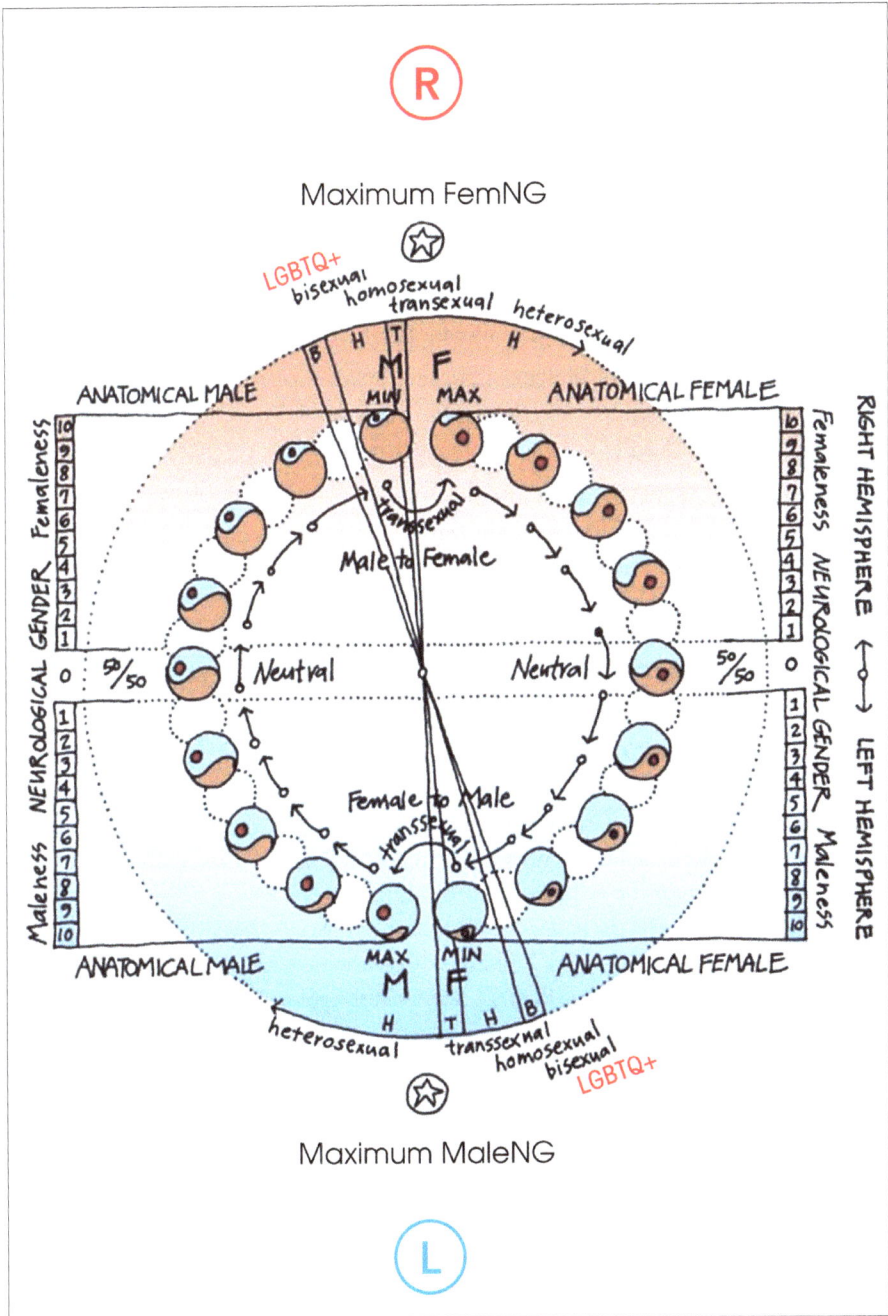

Figure 27. Sexual Attraction/ SOCIAL ORDER: A Speculation

Why, from an evolutionary perspective, would it be advantageous to have a recurring percentage of the population who are non-binary, or who forego having children? The three high-femaleness males cited as exemplars of artistic achievement in Western Civilization: Michelangelo, Leonardo da Vinci, and Botticelli never married and never had children. A significant proportion of the US and European Union (15-18%) do not have children, suggesting that this results in a more resilient population. The wide variations in maleness and femaleness expression, associated talents, and ways of being in the world are shown in *Figure 28* which is a speculation, but consistent with the Pre-human Primate and McGilchrist predispositions. Three Worldviews are shown, ranging between the extremes of Femaleness-to-Maleness: those who choose to give up their life for others, and those who choose to take the life of others—MAXIMUM EMPATHY to MINIMUM EMPATHY.

WORLDVIEW/POLITICAL: Point-of-View
Males and Females on the *femaleness* end of the spectrum tend toward Trust, Innovation, and Flexibility. They range toward:
LIBERAL-REVOLUTIONARY-SOCIALIST thought.
Males and females on the maleness end of their spectrums tend toward Defending the Status Quo, Distrust, and Inflexibility. They range toward: CONSERVATIVE-REACTIONARY-FASCIST thought.

WORLDVIEW/OCCUPATIONS: Ways-of-Being
Males and females on the *femaleness* end of the spectrum tend toward the HUMANITIES, Art, Architecture, Music, Literature, Style, Film, and the Performing Arts. Genius occurs in MUSIC & ART and selfless SAINTS & HEROES are near the extreme.
Males and females on the *maleness* end of the spectrum tend toward SCIENCE and Engineering, FINANCE, Real Estate, Construction, and the MILITARY. Genius occurs in MATH & PHYSICS and selfish egoists, SOCIOPATHS & PSYCHOPATHS, are near the extreme.

WORLDVIEW/ELECTIONS: Leadership
US National Elections reflect fixed Conservative-Maleness 40% and Liberal-Femaleness 40% voter blocks. In the modern era, there were two so-called landslide elections. (1) Johnson-Goldwater 60% to 40% LIBERAL (1964), and (2) Nixon-McGovern 60% to 40% CONSERVATIVE (1972)
With near-fixed mindsets of 40% Liberal and Conservative, only the remaining 20% of independents are persuadable. No 90/10, No 80/20, No 70/30 landslides...only the maximum victory potential of 60/40.

In this Chapter we put forth a new Theory of Genomic Imprinting. We also speculated about how this might inform Sexual Attraction and form Social Order. In Chapter Four we will examine a revolutionary event: the instinctive/subconscious RNA MIND evolving to the overlay of deliberative/consciousness.

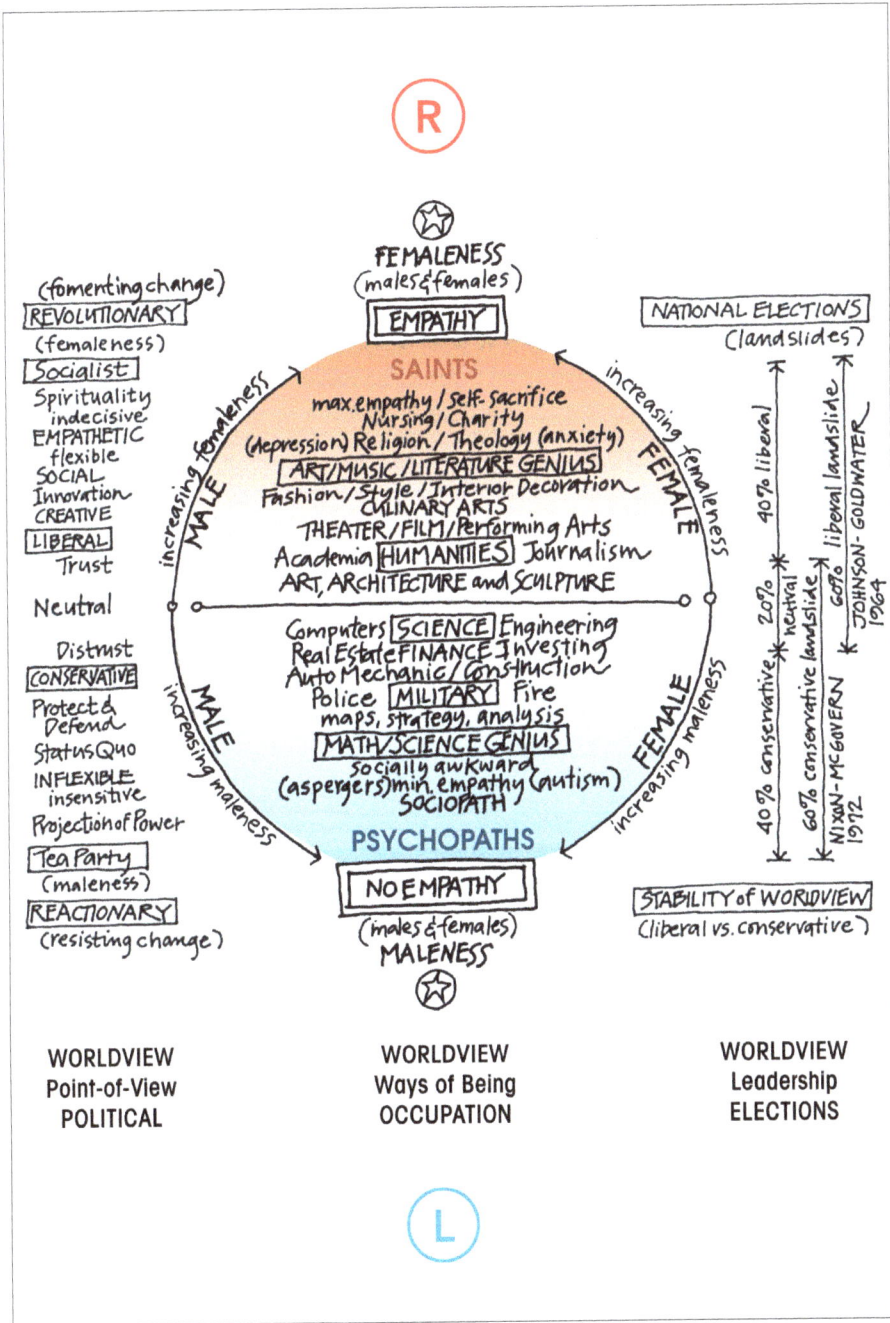

(Ⓡ)

✪

FEMALENESS
(males & females)

(fomenting change)
REVOLUTIONARY
(femaleness)
Socialist
Spirituality
indecisive
EMPATHETIC
flexible
SOCIAL
Innovation
CREATIVE
LIBERAL
Trust

Neutral

Distrust
CONSERVATIVE
Protect &
Defend
Status Quo
INFLEXIBLE
insensitive
Projection of Power
Tea Party
(maleness)
REACTIONARY
(resisting change)

EMPATHY

SAINTS
max. empathy / self-sacrifice
Nursing / Charity
(depression) Religion / Theology (anxiety)
ART / MUSIC / LITERATURE GENIUS
Fashion / Style / Interior Decoration
CULINARY ARTS
THEATER / FILM / Performing Arts
Academia HUMANITIES Journalism
ART, ARCHITECTURE and SCULPTURE

Computers SCIENCE Engineering
Real Estate FINANCE Investing
Auto Mechanic / Construction
Police MILITARY Fire
maps, strategy, analysis
MATH / SCIENCE GENIUS
socially awkward
(aspergers) min. empathy (autism)
SOCIOPATH

PSYCHOPATHS

NO EMPATHY

(males & females)
MALENESS
✪

increasing femaleness
MALE
increasing femaleness
FEMALE
increasing maleness
MALE
increasing maleness
FEMALE

NATIONAL ELECTIONS
(landslides)

40% liberal
60% liberal landslide
JOHNSON - GOLDWATER
1964

20% neutral
40% conservative
60% conservative landslide
NIXON - McGOVERN
1972

STABILITY of WORLDVIEW
(liberal vs. conservative)

WORLDVIEW	WORLDVIEW	WORLDVIEW
Point-of-View	Ways of Being	Leadership
POLITICAL	OCCUPATION	ELECTIONS

(Ⓛ)

Figure 28. Worldview/SOCIAL ORDER: A Speculation

CHAPTER FOUR

HUMANKIND

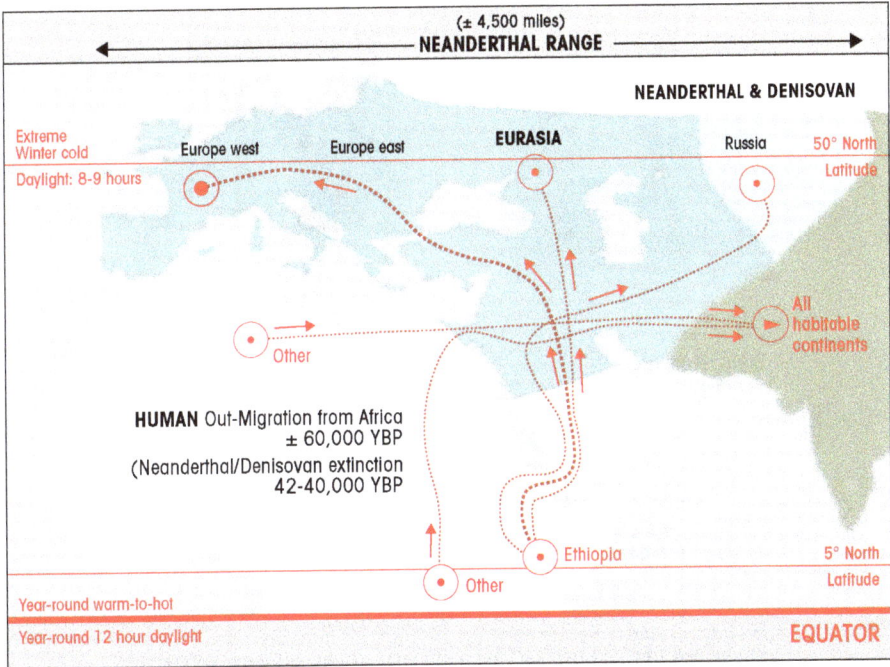

Figure 29. Human Displacement of Pre-humans 50-40,000 YBP

Our closest relative to have walked the face of the Earth is *Homo neander-thalensis*; we were both members of the larger Pre-human family, the genus *Homo*. Neanderthals emerged ±400,000 YBP in Eurasia, and *Homo sapiens* ±300,000 YBP in Africa from a common ancestor: *Homo heidelbergensis*. For over 300,000 years Neanderthals and their sister group, *Homo denisovan*, dominated the largest landmass on Earth: Eurasia. We are direct descendants in the line of *Homo sapiens*, who evolved transitional (*pre-conscious*) Minds over the period 300,000 to 100,00 YBP in Africa—our bridge to consciousness. The evolution to the *deliberative-conscious* Mind 100,000 years ago is the defining trait of Humans and is evident in the archaeological and biological record. For clarity, references to our ancestors, from 100,000 YBP to the present day will be to *Humans*.

The out-migration of Humans ±60,000 YBP (See *Figure 29*) and the following extinction of Neanderthals and all members of the genus *Homo*, established the modern order of life on Earth. All previous out-migrations by members of the genus *Homo* ultimately failed. The emergence of the deliberative-conscious Mind marks the beginning of the Anthropocene, the geological era in which Humans took over—not just all habitable continents, but, through their newly-evolved deliberative consciousness and behavior, the future course of climate change, the water cycle, and species extinction.

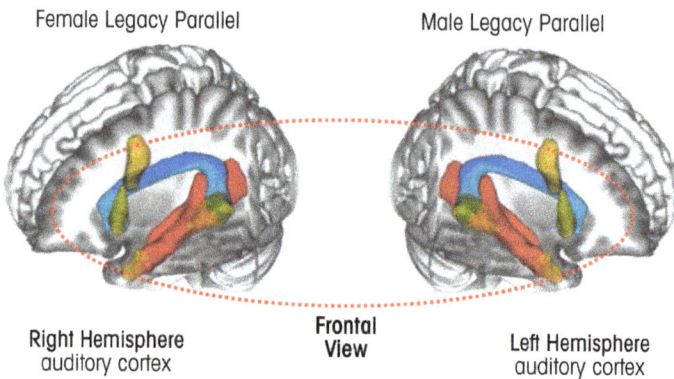

The Binary Origin of Consciousness

Female Legacy Parallel   Male Legacy Parallel

Right Hemisphere
auditory cortex
    Frontal
View
    Left Hemisphere
auditory cortex

Daniela Sammler et al.
"Dorsal and Ventral Pathways for Prosody"[1]

"Here we provide evidence for a multi-stream architecture...
and propose a functional division of labor that parallels
prevailing language models, but in the right hemisphere."

Cell **25** Issue 23 (December 2015) p. 3079-3085
doi: 10.1016/i.cub.2015.10.009

Figure 30. The INFRASTRUCTURE of the INNER VOICE

ASSERTION: Co·GENESIS describes the evolutionary construct within which biological transitions from the *instinctive-subconscious minds* of animals, to the *pre-conscious mind* of Homo sapiens, to the overlay of the *deliberative-conscious mind* of Humans was possible.

Co·GENESIS and the Instinctive-subconscious Mind: Stage One. The expression of the left/right (asymmetrical) Mind via sexual reproduction and RNA-imprinted genes can be thought of as the first step: the variation of the minds of offspring for the resilience of species. This evolutionary order spanned a billion years and produced millions of species. However, the Mind of the individual is instinctive-subconscious and inflexible for life. Evolution of the Mind, like the Body, advanced slowly via Natural Selection: weeding out the least successful and expanding the most advantageous, generation-by-generation.

Co·GENESIS and the Deliberative-conscious Mind: Stage Two evolved with the increasingly interwoven informational flows between the hemispheres, reflected in the recently-discovered binary structures of speech.[1] See *Figure 30*. Auditory and speech capacity of Neanderthals and Denisovans is now well-supported by genetic and anatomical evidence. However, "cross-talk," the deliberation between the hemispheres, our inner voice and consciousness, evolved solely in the line of *Homo sapiens*, the beginning of a new order of life at planetary scale.

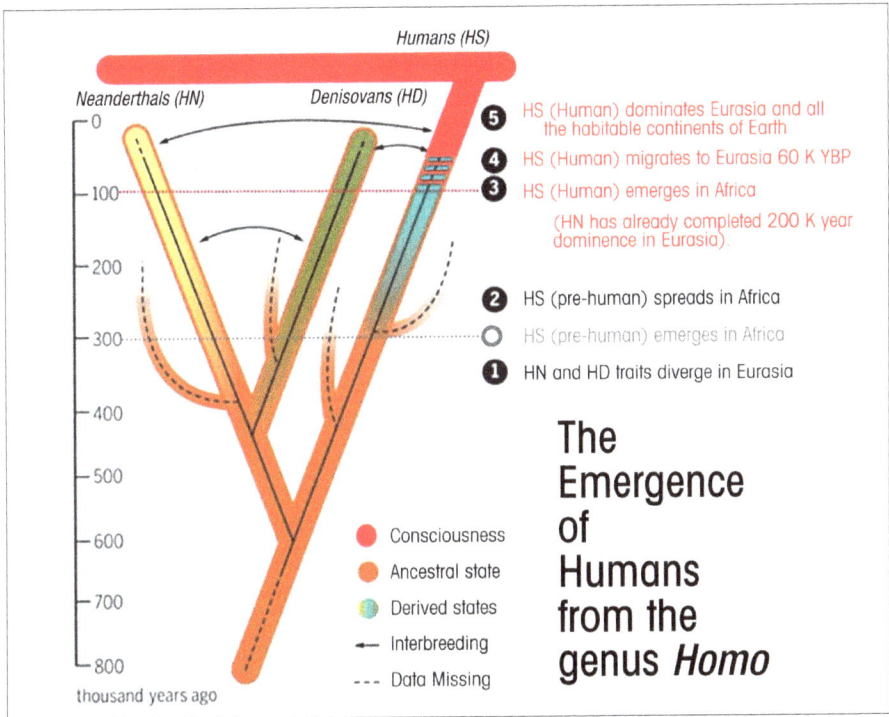

Figure 31. The GLOBAL EXTINCTION of the genus *Homo*

Research teams, in 2020 and 2021, clarified the evolution of speech in the genus *Homo*. Mercedes Conde-Valverde et al. in their article for *NATURE ecology & evolution* (2021) "Neanderthals and *Homo sapiens* had similar auditory and speech capacities"[2] place the auditory bandwidth for vocal communication quite late (after *Heidelbergensis*). Importantly, David Gokhman et al., reported in *Nature Communications* (2020)[3] on a period of intense change of the genes of the facial anatomy and vocal tract in *Homo sapiens*. The anatomy of the larynx, when compared to our modern phonic range, is found to be undeveloped in Neanderthals and Denisovans. In short, all three key players in *Figure 31* had the capacity for language but only Humans had the vocal complexity that would mark an evolution from the instinctive-subconscious to the overlay of the deliberative-conscious Mind. We will examine further biological, archaeological, and cultural evidence supporting this critical transition.

The family tree of pre-humans and Humans was clarified by Julia Galway-Witham and Chris Stringer in "*How did* Homo sapiens *evolve?*" (published in *SCIENCE*, June 2018).[4] See *Figure 31*, an expanded interpretation of the researchers' *species diagram* consistent with the pre-human/Human distinctions laid out in Chapter Three. Red designates the recent emergence of deliberative consciousness in Humans (for the genus *Homo*, an extinction-level event).

Neanderthal dominance of Eurasia, was characterized by their refinement of a distinctive method of shaping stones for tools (knapping), classified as *Mousterian*. Consistently found with Neanderthal remains, this technique is a defining feature of the Middle Paleolithic era. Such a lengthy period of domination with only incremental modification of stone tools and weapons indicates a species with a highly successful physical and behavioral adaptation to its environment; there was little natural selection pressure for further change or innovation. A parallel in the animal kingdom is the shark: so successful in its competitive niche that it changed little over hundreds of millions of years.

While Neanderthals maintained dominance of Eurasia, *Homo Sapiens* and later Humans took on the established African bands of *Homo heidelbergensis, Homo naledi* and, most likely, other descendants of the genus *Homo*. Harvard's Mary E. Prendergast and David Reich supervised an international team headed by Mark Lipson in completing four genome-wide DNA analyses from which they inferred an Africa-wide family tree of Human emergence, published in NATURE (January 2020).[5] Two of the four lineages are of particular interest to us and, in contrast to all others in the genus *Homo*, provide evidence of modern behavior ±100,000 YBP in Africa. South Africa's Blombos Cave and Diepkloof Rock are the sources of archaeological evidence for the southern lineage, and East Africa's Porc-Epic Cave and Aduma (Middle Awash River Basin), Ethiopia, are the sources for the critical eastern lineage.

What do researchers mean when they say *modern behavior*? Two examples will suffice for our purposes: *symbolic thought* (the projection of information from the mind into the external world in representational form as in drawing, painting, sculpture) and *innovation* (going beyond the reassembly of "found" materials/phenomena to devise complex methods exceeding naturally-occurring phenomena). Both behaviors, I argue, are the product of conscious deliberation between the male and female legacy hemispheres and occur only in Humans.

A well-documented failure of out-migration by *Homo sapiens* ±194,000-220,000 YBP in the Levant (modern day Israel), presents evidence of pre-human interbreeding between early *Homo sapiens* and Neanderthals. However, when Humans and Neanderthals came face-to-face in Eurasia ±160,000 years later, the biological gap had expanded—they could no longer successfully interbreed, as reported by Sankararaman et al. in *NATURE* in 2014.[6] The two apex predators came face-to-face in Neanderthal's homeland of Eurasia, most notably in the lands we now call Western Europe, along the 50° North Latitude line (the range of woolly mammoths) ±50-42,000 YBP. See *Figure 29*, page 53.

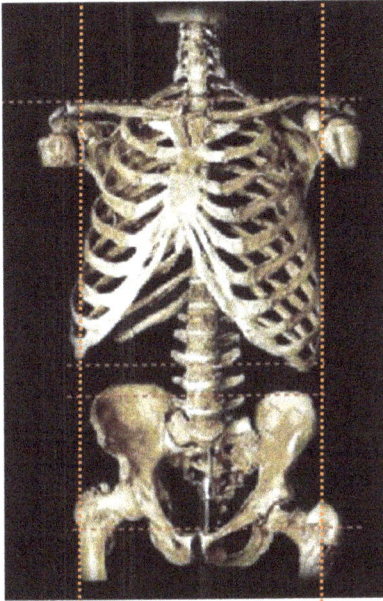

Homo neanderthalensis (HN)
HN has a wide stance for overturn
resistance in face-to-face combat.
HN skeleton reveals impact resistance, heavy
musculature, mid-section protection, low heat loss,
slower body movement.

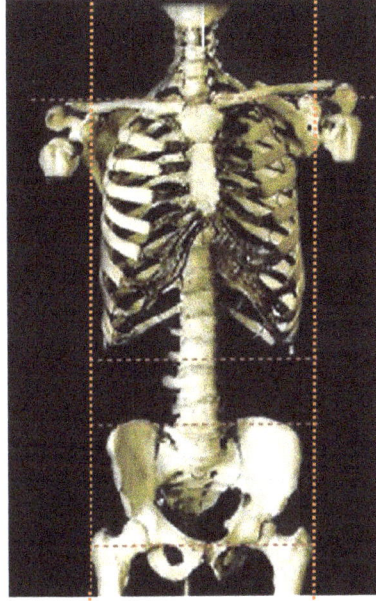

Homo sapiens (HS)
HS is slender for speed, endurance, and
mobility, has high heat loss. HS skeleton
has extended spine with rotational flexibility for
launching projectile weapons, creating
mid-section exposure of vital organs.

Figure 32. NEANDERTHAL and HUMAN Body Cores

In addition to different mindsets, their physical characteristics were strikingly different. See *Figure 32*. Neanderthal musculature meant they could wield heavy stone weapons (thrusting spear, hand axe) in close combat and, given their significantly lower center of gravity and wide stance, they could quickly get Humans (or any other member of the genus *Homo*) off their feet. Humans were in no way physically prepared for such a face-to-face confrontation, having a slender build, high center of gravity, exposed mid-section, and narrow stance. But Humans had figured out how to project deadly force from a distance to avoid such direct physical contact, and how to field greater numbers at the point of contact. While Neanderthals in Eurasia had evolved superior brute strength, tolerance of the cold, and night vision for these dark North Latitudes, Humans had evolved their ultimate weapons in the neurological realm of deliberative consciousness: innovation, anticipation, and strategic thinking, which overcame the long list of their physical disadvantages.

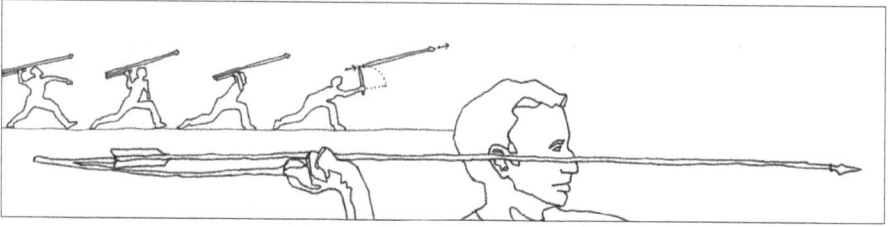

Figure 33. Complex Projectiles: Higher Velocity, Range, and Accuracy

John J. Shea and Matthew L. Sisk, writing in *Paleo Anthropology* in 2010,[7] proposed that complex projectile weapons such as the bow and arrow and/or the atlatl (large arrow) with small stone tips were the key innovations by Humans that enabled their successful Out-of-Africa migration. See *Figure 33*.

These smaller stone tips are found with Human, but not Neanderthal remains. Yonatan Sahle and Allison S. Brooks in 2019 authored an updated assessment of this transition in *PLoS ONE* (May 2019)[8] stating: "At Aduma (Middle Awash, Ethiopia), morphometric, hafting, and impact damage patterns in several lithic [stone] point assemblages suggest a shift from simple spear technologies [thrusting and/or hand-cast] to complex projectiles." The time period of the shift is ±100,000-80,000 YBP, and the atlatl is identified as the likely first complex projectile mechanism. The low sea levels of ±60,000 YBP likely facilitated the out-migration of Humans who could more easily cross the Red Sea's newly-narrowed neck to Yemen and Eurasia beyond.

Why were complex projectiles so significant given that *Homo heidelbergensis* utilized hand-cast wooden spears over 400,000 YBP and Neanderthals wielded thrusting, stone-pointed spears over 300,000 YBP? The answer: those earlier weapons were limited by the strength of whoever was thrusting or throwing them—the metrics of Natural Selection. The release velocity of spears thrown with accuracy was 30 to 35 mph; prey included the Woolly Mammoth (20 mph), Rhinoceros (30 mph) and the Bison (35 mph). In contrast, Humans used the *mechanical advantage of the lever* in the atlatl to reach velocities of 45 to 55 mph with greater accuracy and distance (safer). Birds in flight, deer, and horses became more viable as prey. Importantly arrows, unlike spears, could not be effectively thrown back by the Neanderthals. The human projection of deadly force in ways not present in Nature had begun (arrows to bullets to bombs to missiles to drones).

Neanderthals followed the path of Natural Selection, becoming the most robust and fearsome predator of the genus *Homo*. Their skulls bulged with an expanded and frightening capacity: low-light or night vision; they were the ultimate stalkers of unsuspecting Humans and animal prey as calculated by Pearce, Stinger and Dunbar et al. See *Figure 34*.[9]

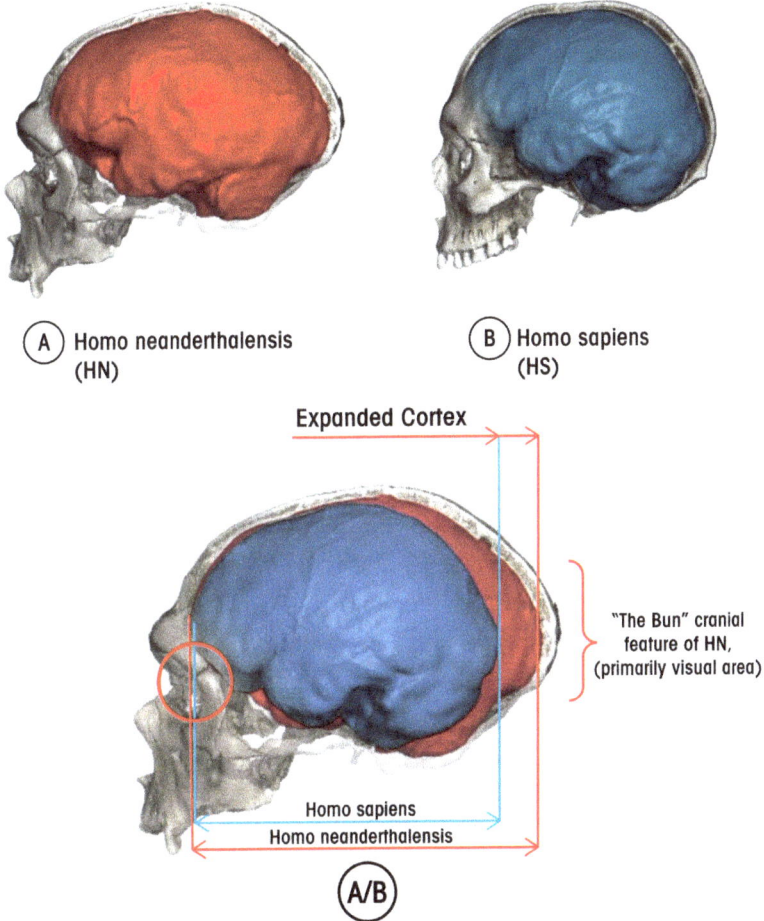

(A) Homo neanderthalensis (HN)

(B) Homo sapiens (HS)

Expanded Cortex

"The Bun" cranial feature of HN, (primarily visual area)

Homo sapiens
Homo neanderthalensis

(A/B)

Superimposed HS Hemisphere on HN: Red = Cortex Adaptation/Expansion

HN Ocular Mass = 34.15 cm^3 while HS = 29.51 therefore HN = 14.7% greater
HN Ocular Area = 1,403.98 mm^2 while HS = 1.223.39 mm^2 therefore HN = 15.72% greater
HN Cerebral Mass resulting from Expanded/Adapted Cortex: HN (approx.) = 15.0% greater

NOTE: Ocular Mass and Ocular Area comparisons are from: Pearce, Stringer and Dunbar et al, "New insights into differences in brain organization between Neanderthals and anatomically modern humans," the *Proceedings of the Royal Society of London. Series B,* Biological sciences, (2013) doi: 10.1098/rspb.2013.0168[9]

Figure 34. NEANDERTHAL ADAPTATIONS to 50° North Latitude

Complex projectiles were just one indicator of the Human competitive advantage. Genetic, archaeological, and skeletal evidence confirm that Neanderthals were organized in small groups or "bands" of closely related members (parents, mates, siblings, aunts, uncles, cousins, etc.) of 5 to 15 members that were isolated and thinly settled across Eurasia for hundreds of thousands of years. In a high threat environment, this level of connectivity—the chemical/hormonal bonding of sexual relationships and parent/child meant that the central test for survival of the band was met—the willingness to fight to the death in defense of the group. Over this vast time frame there is little evidence of increased Neanderthal complexity in artifacts, socialization, or expanded group size. By contrast, the arriving Humans quickly advanced to larger groups and socialization as seen in their higher genetic diversity versus Neanderthal's low diversity and evidence of in-breeding. I have long believed that each perceived a different world; that their behaviors reveal a fundamentally different stage in development of the binary RNA MIND. To my amazement, the earlier-mentioned Sankararaman et al. team in "The Genomic Landscape of Neanderthal Ancestry in Present-day Humans" NATURE (2014)[6] pointed out this difference. See Figure 35.

The Sankararaman team, under Harvard University's David Reich, was focused on interbreeding between Humans and Neanderthals as evidenced by the modern genetic remnants of such unions, and in their search, examined the genomes of 1,004 present-day Humans. In a process similar to reverse-engineering, they were able to identify traits and tissues which were the most incompatible with interbreeding (most different) in Neanderthals. In observations beyond the center of their research, the team mentioned that Neanderthals were "depleted" in an area that is conserved in Humans: RNA processing. After studying 101 pages of Supplementary Information, I found that miRNA, tRNA and the endocrine system's adrenal gland, key elements of deliberative consciousness in Humans, were depleted in Neanderthals.

Why did Humans advance to large groupings within the same hostile setting that so challenged Neanderthals? Humans simply perceived a different world. Neanderthals had unique predispositions at birth consistent with Co·GENESIS and the RNA Mind, but the decision-maker in real time was instinctive subconscious. A Human instinctive subconscious was, and still is, present but can be over-ruled by our *deliberative-consciousness*. This breakthrough created *hyper-perception*—we see the world with us in it: self-awareness—*the perceptual triangulation of two minds (here) and the object observed (there)*. Humans were intrigued by the world they could suddenly see, and the disturbing questions it raised: How was this world created, and by whom? Why am I here?

Neanderthals (HN)

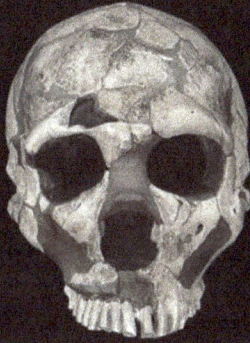

"Neanderthal ancestry is also depleted in conserved pathways such as
RNA processing."
Sriram Sankararaman et al., "The Genomic Landscape of Neanderthal Ancestry
in Present-Day Humans," (2014)
Nature 507 354-357 doi: 10.1038/nature12961.[6]

HN: RNA Binding (DEPLETED) including miRNA and tRNA
HN: RNA Transport (DEPLETED) including tRNA
HN Adrenal gland (at VARIANCE): 2nd lowest HN ancestry of HS tissues
listed in the table of "Low Neanderthal Ancestry"
(key factor in sexual orientation and gender identity)

Humans (HS)

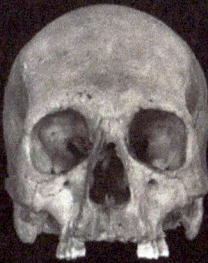

HS: RNA Binding: miRNA and tRNA (CONSERVED)
HS: RNA Transport tRNA (CONSERVED)
HS: Adrenal gland (CONSERVED)

Figure 35. RNA DIFFERENTIATION in Neanderthals and Humans

The response of the deliberative-conscious Mind to these frightening questions was, and is, the explanatory narrative: the story of creation, mortality, and the
role of Humankind. The centerpiece of *Figure 36* is the Lion Man (A), sculpted
from the ivory tusk of a Woolly Mammoth 40,000 YBP and standing 12.2 inches
high; it is considered to be the oldest evidence of religious belief in the world.
Merging the head of the Eurasian Cave Lion and the body of a Human warrior,
it is a masterpiece of a mythical, supernatural being. The photo and drawing in
Figure 36: F and *G* capture the detail of the left upper torso in profile, the most
well-preserved area. At the moment that Humans were taking over the strong-
holds of the Neanderthal homeland in Western Europe, we have this evidence
of Humans as a belief-bound group. With a *shared belief worth dying for*, Human
groups expanded beyond the blood/sex biological limits of the Neanderthal bands.
Language, the core of the deliberative-conscious Mind, and the power of oratory
would have been essential in order to inspire and unite large groups of unrelated
individuals under a fight-to-the-death belief. Humans, now with command and
control, projectile weapons, and larger numbers, drove Neanderthals, the genus
Homo, and all remotely Human-looking species from the face of the Earth.

The Lion Man, an exemplar of sculpture, suggests that the two-dimensional
precursors, drawing and painting, were mastered much earlier. Abstract drawing
at Blombos Cave, 73,000 YBP,[10] has been found, as well as engraved bands on
ostrich eggs (used as water containers) at Diepkloof Rock Shelter, dated 60,000
YBP.[11] The East Africa lineage site of Porc-Epic Cave yielded 21 ochre-processing
tools and powder colors dated to 40,000 YBP.[12] Collectively these are the cultural
signature of Humankind: the projection of representational expression from the
deliberative-conscious Mind into the World.

Neanderthals had no record of what they simply could not observe, no
comparable representational drawing, painting, or sculpture. The use of red ochre
by Neanderthals for body paint, perforated shells for body ornamentation, and
stone tipped spears for hunting are parallels to chimpanzees stripping sticks to
fish for ants and termites, sea otters breaking open shellfish with rocks, or the
octopus's assemblage of shells to create an enclosing body camouflage and escape
vehicle. All such arrangements, applications, and repurposing of found objects
and phenomena (even the use of fire) are simple adaptive behaviors. Nature is
also replete with hyper-complex inherited survival and reproductive behavior
via Co·GENESIS (i.e., the bower bird's nest and dance, or the bee's social hier-
archy and hive). However, none of these confirm the existence of deliberative-
consciousness, which has occurred only once—in Humankind.

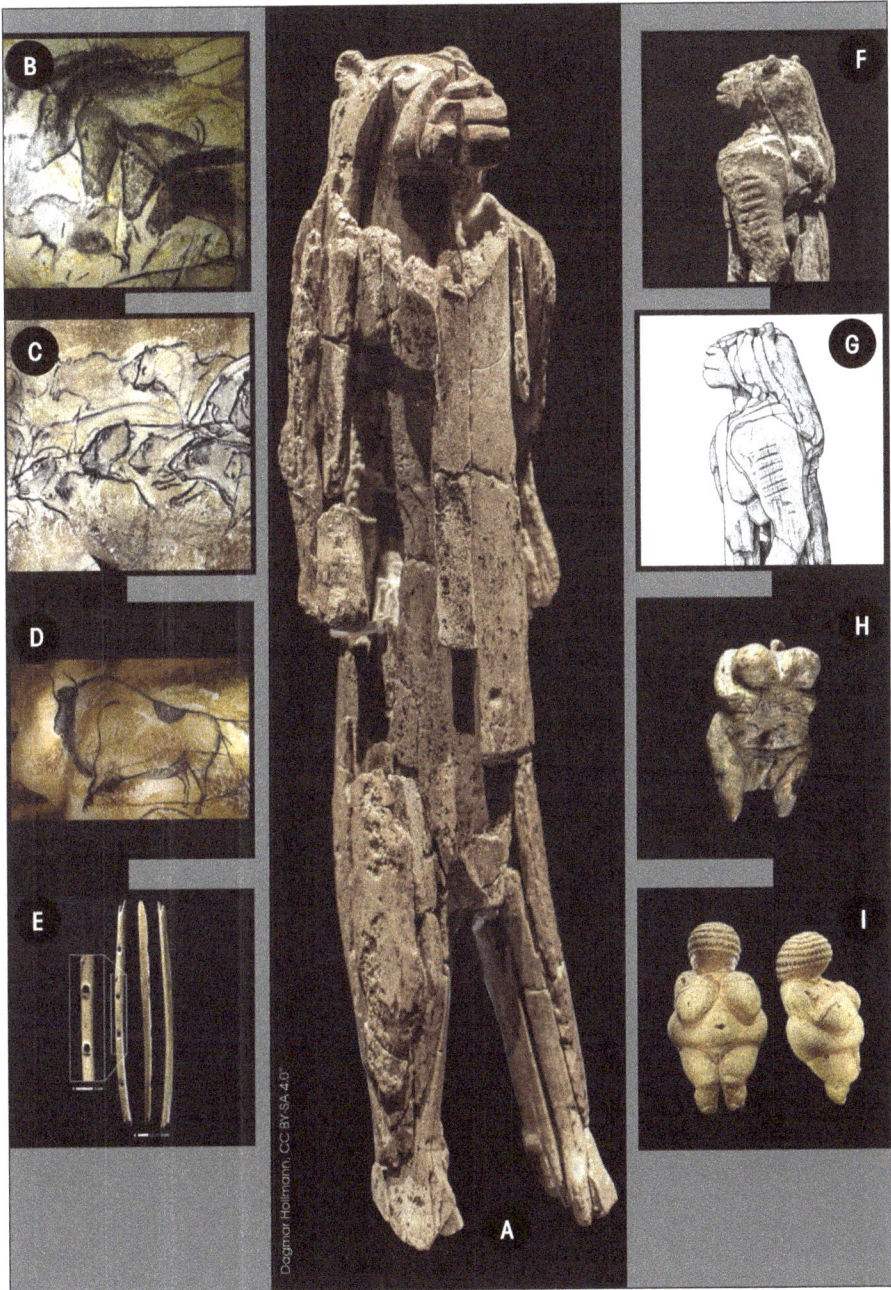

Figure 36. Deliberative-Conscious Projections of the Human Mind

NOTE: A, B, C, D and E (a vulture wing bone carved to form a flute) are Human works just prior to the extinction of Neanderthals in Europe. Female figurines H and I, 40,000 YBP and 30,000 YBP, respectively, bracket the extinction period of the Neanderthals through to the last hold-outs in the sea-side caves of Gibraltar, Spain.

Figure 37a. Hall of the Bulls at Lascaux Cave, France (left segment)

The significance of what happened in Africa was the break-out by Humans from the RNA MIND's billion-year linear evolution of information at the speed of one generation to the next—instinctive and subconscious—to the near-instantaneous evolution of observed information by the Mind in conscious deliberation. Most importantly, *co-evolution* with collaborating and competing Human Minds further accellerates the pace of evolution. An exemplar of such radical acceleration is the advent of powered flight. The Wright brothers were first, on December 17, 1903 at Kitty Hawk, North Carolina (flying for 59 seconds, traveling 852 feet). On July 20, 1969, Neil Armstrong stepped onto the surface of the moon, achieving final victory in a twenty-year fierce competition with the USSR. For us, the 300,000 years Neanderthals spent refining a stone-shaping technique is nearly impossible to imagine.

The Human Mind (for better and worse) was empowered to stand outside the Environment, able to perceive and evolve the Environment to serve itself—rather than being evolved by the Environment. As I sit here at my desk, I am surrounded by hundreds of objects adapted to the Human body: the keys I am typing on are cupped to my fingertips; pens fit perfectly to hand, the chair to body, the computer screen to field of view, etc. Evolutionary feedback from our environment of buildings, cars, trains, and planes cries out: "This body is perfect, change nothing!" By contrast, the environment at 50° North Latitude faced by the Neanderthals, left nothing of their physical bodies unchanged.

Figure 37b. Hall of the Bulls at Lascaux Cave, France (right segment)

In this Chapter, we have followed the emergence of Humankind which one day, in my opinion, will be defined as a separate species (*monotypic genus*) because of its defining differentiation: deliberative consciousness, language and the inner voice—the inseparable three-part basis for the Human rise to global primacy. The competition between Neanderthals, the apex predator of the old order, and Humans, the apex predator of the new order, left biological and physical evidence from which we have reconstructed the two very different worlds they perceived. The baseline observations, further elaborated in Chapters 5 and 6, follow:

1) In the period 300,000 to 100,000 YBP, the brain's parallel structures of speech began to interact in *Homo sapiens* in real-time, leading to the evolution in Humans of language, the inner voice and their expression: the deliberative-conscious Mind.

2) The timing and context of this revolutionary event is knowable based on the speed of innovation and projection of representational artistry never before seen in any life form. See *Figure 37a and 37b*.

3) Deliberative consciousness requires the speed and volume of data processing found in high level RNA function (these functions have been found to be conserved in Humans, and depleted in Neanderthals).

4) The emergence of deliberative consciousness raised unnerving questions: Who created the Universe? Why am I here? What happens when I die? The answers, by the Humans facing the Neanderthals, inspired unrelated groupings to unite, to rise above their fear of death, and to take and hold the lands of Eurasia and the Earth beyond.

CHAPTER FIVE

THE MECHANISMS

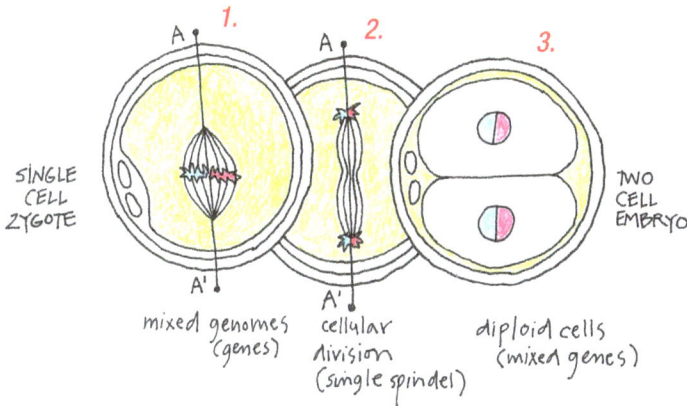

Figure 38: First Cellular Division after Fertilization

In Chapter 4, we noted that RNA processing, miRNA, and tRNA, were found by the Reich-Sankararaman team to be conserved in Humans and depleted in Neanderthals. Importantly, all three functions are essential to the information processing and pathways of the deliberative-conscious Mind. In this chapter we return to the molecular level to understand how the RNA MIND records adaptive characteristics over a lifetime and updates the ancestral male and female legacies as they are being passed to offspring. The biological mechanisms identified here build on Chapter 3, which was foundational in my rejection of two of the tenets of biology:

1) **Weismann Barrier**: the assertion that no information from the parental body cells (soma) passes into egg and sperm (therefore life experiences cannot pass down).

2) **Mixing of Genes**: the assertion that parental RNA and DNA are randomly mixed in the first cellular division, preventing life experience adaptations from passing down (This critique was proffered by a molecular biologist who reviewed *A Convergence of Two Minds* in 2015).

Biology and the Modern Synthesis are based on an immediate mixing of genes as simple as 1, 2, 3 in *Figure 38* above. To the contrary, the Co·GENESIS assertion is that the two Codes of Life are reconciled, updated, and re-expressed by RNA in multiple steps during the transit to the womb: fertilization-to-implantation. For purposes of illustration, the working assumption is a six-day transit. This Chapter and the following Chapter 6 cover this critical sequence of early cellular development and adaptation with twenty-four illustrations, explained on the facing page. These images are simply described and to the point, to achieve an accessible insight.

BRIEF: 39

If there was one illustration in this book, it would be *Figure 39*, showing the maintained separation of the male and female halves of the genome after the joining of egg and sperm to form the zygote. The Theory of Co·GENESIS requires the separate integrity of male and female information across the generations, a phenomenon I was repeatedly assured by molecular biologists did not exist. When I came upon the microscopy slides and article by the EMBL research team of Ellenberg-Reichmann et al. in the 13 July 2018 issue of *SCIENCE*,[1] I knew that my three-year quest to find a biological mechanism for the separate pathways of male and female RNA at the beginning of life had been rewarded. This separation creates a bridge for the RNA MIND to enter the next generation.

I wanted to shout "Eureka!" only to find even greater relevance to my quest in the observations of Agata Zielinska and Melina Schuh of the Max Planck Institute for Biophysical Chemistry reviewing the Reichmann findings in the *Insights/Perspective* section of the same issue of *SCIENCE*:[2]

"Whether the spatial separation of parental chromosomes (genes) has any advantages for the developing mammalian embryo *is unclear.*"

[Author's emphasis]

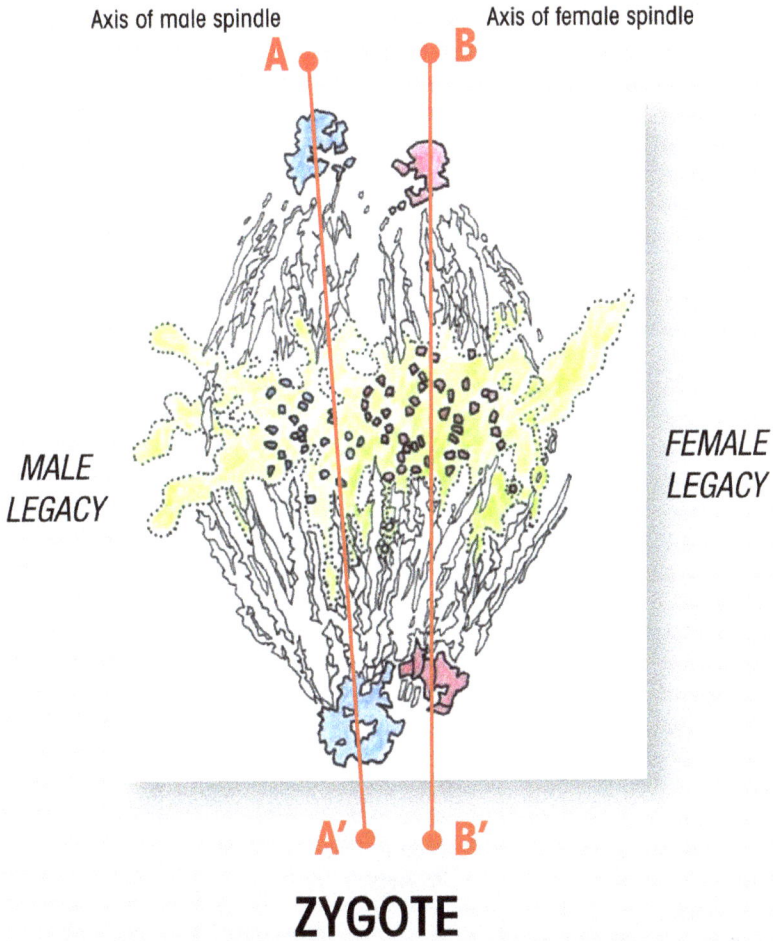

Axis of male spindle   Axis of female spindle

A   B

MALE LEGACY

FEMALE LEGACY

A'   B'

ZYGOTE

Reichmann et al., "Dual-spindle formation in zygotes keeps parental genomes apart in early mammalian embryos" *Science* **361**, 189-193 (July 2018)[1]

"We used light-sheet microscopy to show that two bipolar spindles form in the zygote and then independently congress the maternal (female) and paternal (male) genomes. These two spindles aligned their poles before anaphase but kept the parental genomes apart during the first cleavage."

Figure 39. DUAL-SPINDLE FORMATION: Male and Female Legacy

BRIEF: 40

By direct comparison, *Figure 40* underlines the open-ended nature of the **RNA Binary**. The Ellenberg-Reichmann team characterized the continued separation of the male and female information (the genomes) as *compartmentalization* at the 2-cell stage, which has been shown to continue on to the 4-cell stage in mouse embryos. The team also reported a similar delay (until the 8-cell stage) in cattle embryos before the three basic cell-types appear: male (paternal imprinted), female (maternal imprinted) and diploid (mixed). In Humans, I am using the 16-cell stage as the working assumption, the most likely timing for closure of the genomes and arrival of the 3 cell types.

This initial period when the "hood is up" and no mixing is taking place, allows the surrounding small male and female RNA messenger molecules (miRNAs, tRNAs, etc.), to establish communication between the two parental genomes (unrelated ancestral and adaptive legacies). A reconciliation of the two Codes of Life, through a process of genetic (Body) and epigenetic (MIND) synthesis, is a core function of Co·GENESIS and would logically be initiated during the traverse down the fallopian tube: from the first cell, the zygote, to implantation in the womb (uterus).

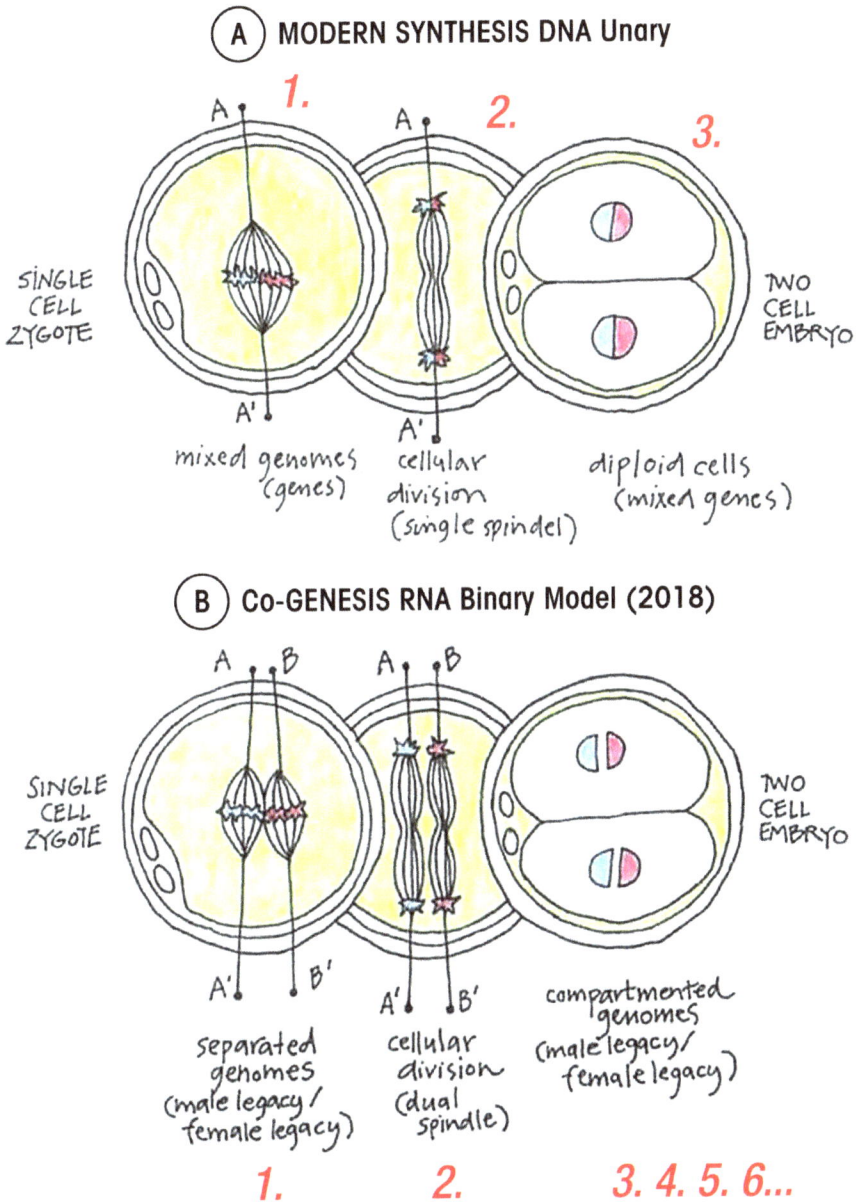

Figure 40. DNA UNARY vs. RNA BINARY and On-going Progressions

BRIEF: 41

By 1868 the existence of cells was well known, although the RNA organic molecules within the cell were not. Nonetheless, Charles Darwin suggested in his *Pangenesis* hypothesis[3] that extracellular particles of information (too small to be seen) flow from the organs of the body, and pass down information to the gametes and on to the embryo. In effect, Darwin put forth a biological mechanism supporting the earlier theories of Jean Baptist Lamarck: that life experiences pass down to subsequent generations. Lamarck's famous example was giraffes evolving long necks to reach high leaf clusters in trees of the African savanna.

Co·GENESIS affirms Darwin/Lamarck's theories of an inheritance of adaptive characteristics via extracellular particles passing down information: the small RNAs: miRNA, tRNA, etc., carrying adaptation updates from the previous generation. The larger concept of RNA male and female genes (ancestral) and the small RNA male and female molecules (parental adaptations) acting jointly to reconcile the binary RNA Mind, is the new territory of Co·GENESIS.

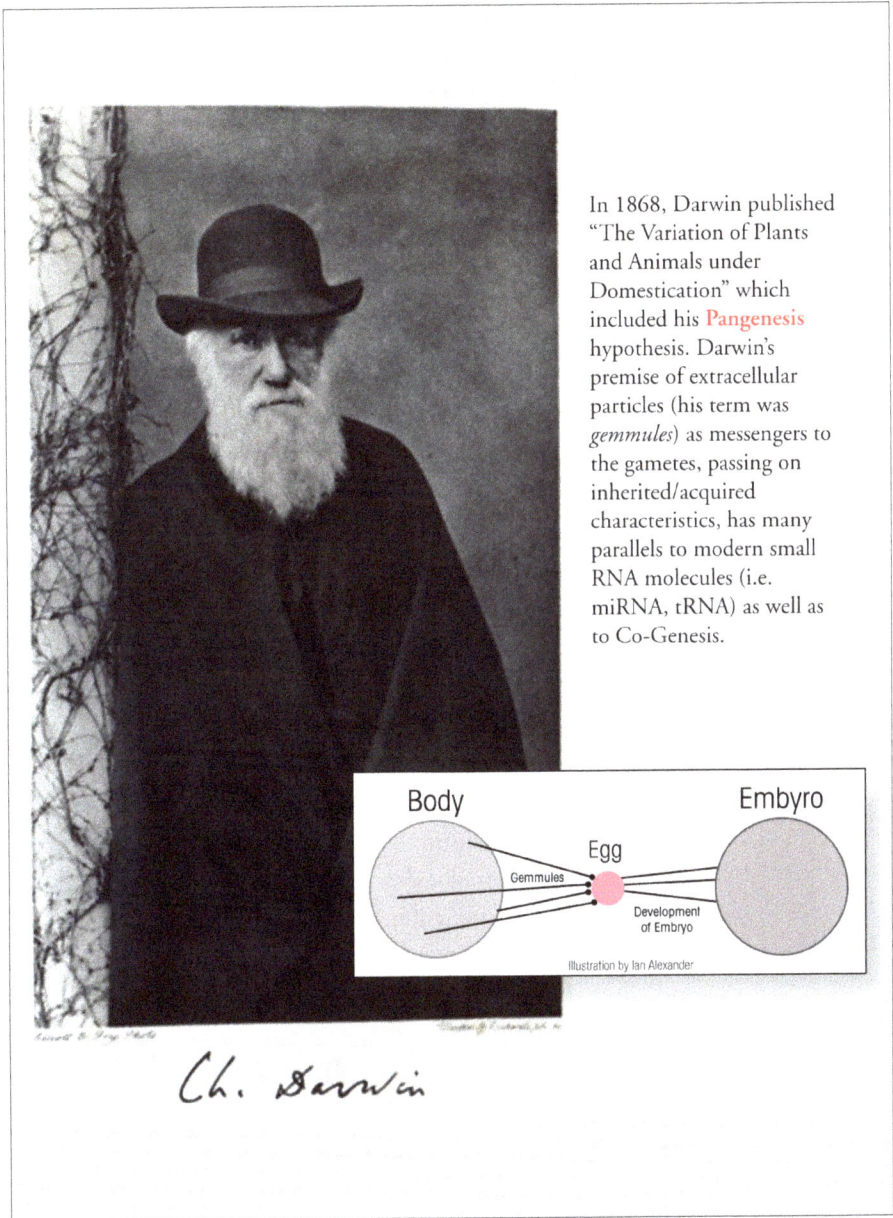

In 1868, Darwin published "The Variation of Plants and Animals under Domestication" which included his Pangenesis hypothesis. Darwin's premise of extracellular particles (his term was *gemmules*) as messengers to the gametes, passing on inherited/acquired characteristics, has many parallels to modern small RNA molecules (i.e. miRNA, tRNA) as well as to Co-Genesis.

Figure 41. Charles Darwin (1809-1882)

BRIEF: 42

Although the upper diagram by Ian Alexander describes Pangenesis, it also resembles Genetics as currently understood. Both envision flows of information into the zygote, resulting in a collision, a mixing of biological information, that establishes the embryo's developmental course.

The lower Co·GENESIS diagram in *Figure 42* uses the same graphic tools to illustrate the more complex formation of the RNA Mind, which directs and regulates the RNA Body over a lifetime. Your paternal and maternal *ancestral minds* are passed down in the form of RNA(m) + RNA(f) imprinted genes together with *adaptive updates* from your parent's environment via miRNA(m) + miRNA(f) small molecules. The reconciled male and female legacies will inform a new Code of Life. Later, at ± 4 months into the gestation period, after an adapted and synthesized Mind is adequately advanced, genomic imprinting transforms the RNA (updated) binary cells back to unary reproductive cells for the fetus: male sperm or female eggs, ancestral legacies which are updated and separate again. Therefore, the unique RNA Mind, the individual Human (able to be passed on) is apparently not fully formed until ±4 months.

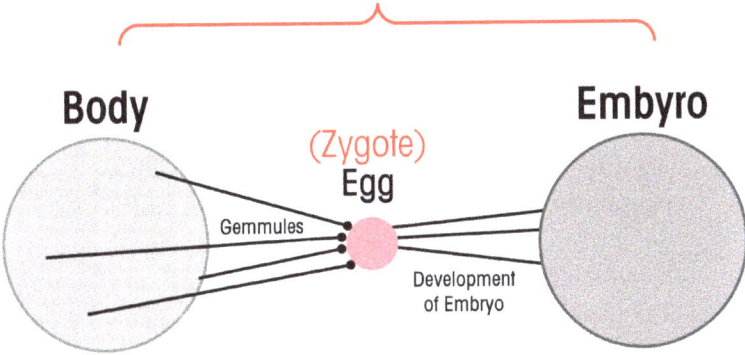

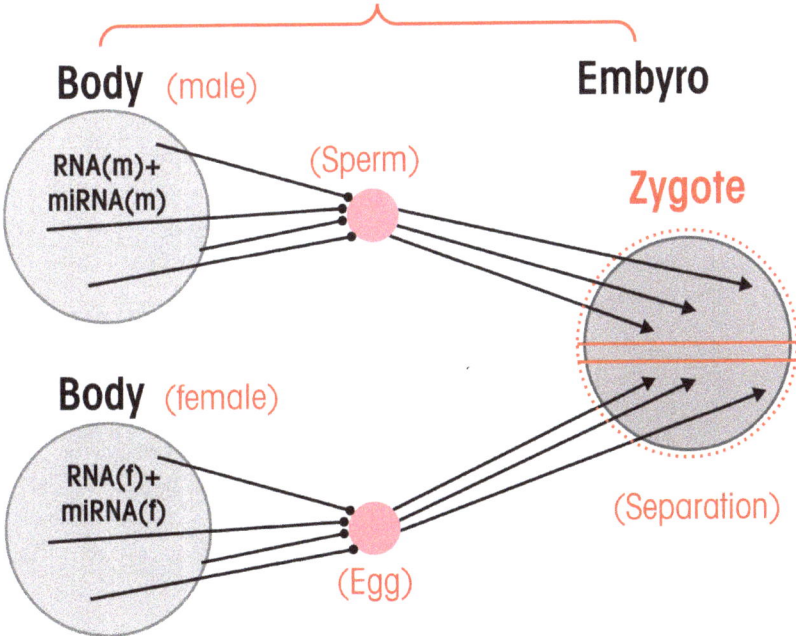

Figure 42. PANGENESIS and Co·GENESIS: Information Flows

BRIEF: 43/44

The nematode (aka roundworm), shown in *Figure 43*, is believed to have evolved nearly a billion years ago as one of the most primitive sexually-reproducing life forms, the first to have a mouth/ digestive track/anus body order. Many of the more than 40,000 species of nematodes have fewer than 1,000 cells and occur in plants and animals (±60 occur in humans only).

Figure 44 compares the nematode miRNA molecule vs. the Human miRNA molecule. The term used to identify the most critical biological elements in an evolving life form is "highly conserved," meaning very little change, even over vast spans of time. Amazingly, the miRNA of the ±37 trillion-celled Human conserves the geometry, complexity, and familiar four-nucleotide assembly of the nematode a billion years later. These near-universal miRNAs are asserted here as the primary intergenerational messengers of Co·GENESIS, carrying down and reconciling acquired male and female life experiences in offspring.

O'Brien et al., writing in *Frontiers of Endocrinology* (2018),[4] confirm that miRNAs circulate in biological fluids throughout the body and they have been located in plasma, serum, spinal fluid, saliva, tears, breast milk, urine, seminal fluid, and ovarian follicular fluid. Sexual intercourse, in this context, is a continuous liquid-to-liquid transfer and commingling of miRNAs from male and female as would be required to reconcile and implement a unique Code of Life.

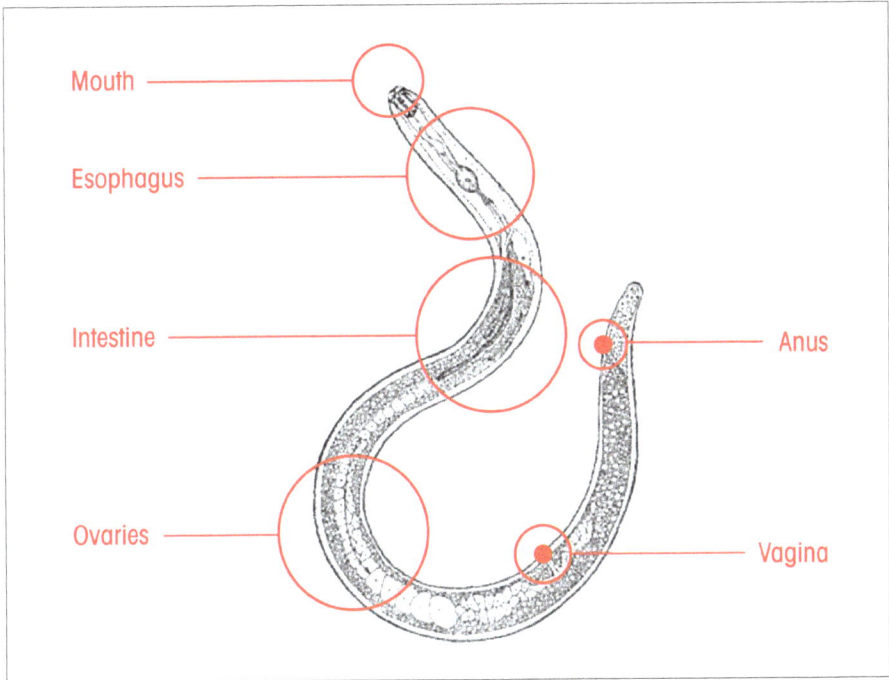

Figure 43. Nematodes, the earth's most populous multicellular animal

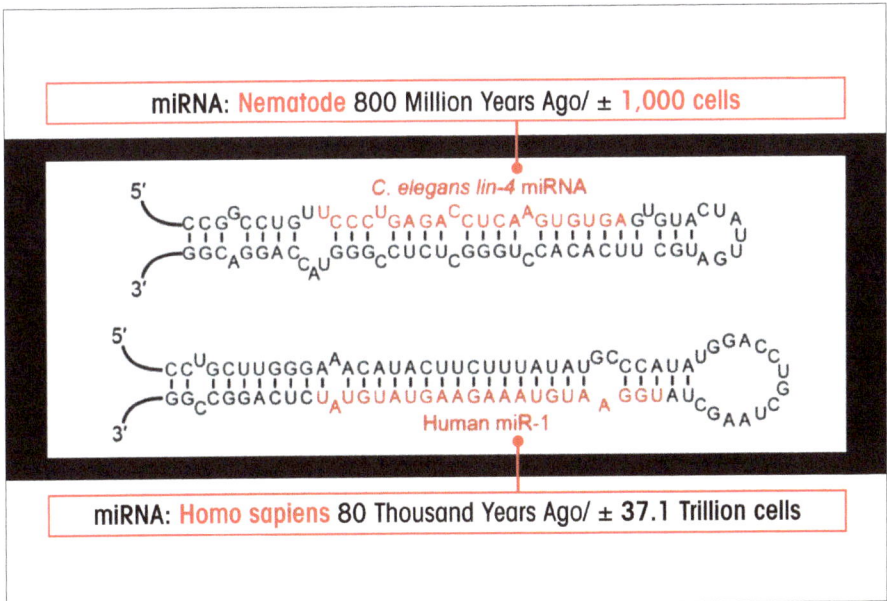

Figure 44. miRNA: Molecular Messengers Generation-to-Generation

BRIEF: 45

Do not be intimidated by this image. There is a simple explanation for the extraordinary event taking place in *Figure 45*. This is a large flow diagram of sperm moving from formation-to-maturation-to-release for a chance to fertilize the female egg. The RNA-imprinted genes are within the sperm along with four types of small RNAs (e.g. miRNA and tRNA) at the start, Point A. However, the main event occurs as sperm passes from **Point A** to **Point B**. During this transit, additional waves (five more species of miRNA and two more species of tRNA) are on-boarded. This last-second surge of information occurs in every projection of sperm over a lifetime—a copy of "today's paper," the latest male adaptive information heading for a merger with the female's current edition. In this manner, the adaptations of the parental lives lived, have a pathway to update ancestral RNA.

The University of Massachusetts Medical School with Oliver J. Rando, leading the team of Sharma et al., publishing in *Developmental Cell* (2018),[5] had the courage to call this out as a "soma-to-germ-line" transfer (a violation of the Weismann Barrier) and, I would argue, another step away from DNA primacy. The male sperm are acting as tiny jump drives, information-packed, on their way back to the female server to download the field report on the male half of life. No such space limitations exist at female headquarters where provisioning for the massive egg and various additions of organs (organelles) for cells-yet-to-be are formed and informed, often with minimal male input.

A **BASIC MiRNA in Sperm**

(miRNA let-7a, miRNA-26a, tRNAf-Glu-CTC, tRNAf-Gly-GCC)
Sperm arrive at the near end (proximal) head of the epididymus **A** from the testes
to begin a maturation sequence **A** to **B**. At this beginning point their small RNA
payload includes the following miRNAs and tRNAs:

(miRNA let-7a, miRNA-26a, tRNAf-Glu-CTC, tRNAf-Gly-GCC)

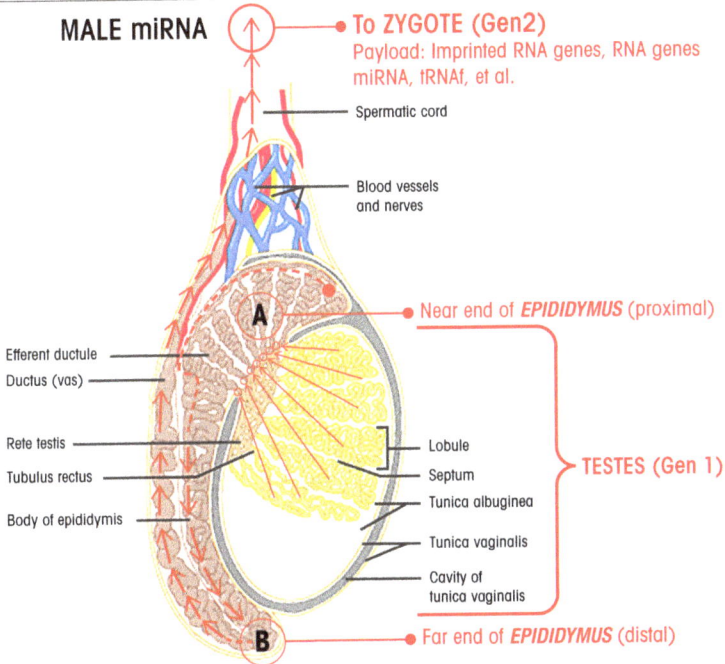

MALE miRNA ↑

To ZYGOTE (Gen2)
Payload: Imprinted RNA genes, RNA genes
miRNA, tRNAf, et al.

Spermatic cord

Blood vessels
and nerves

Near end of *EPIDIDYMUS* (proximal)

Efferent ductule
Ductus (vas)

Rete testis
Tubulus rectus
Body of epididymis

A

Lobule
Septum
Tunica albuginea
Tunica vaginalis
Cavity of
tunica vaginalis

TESTES (Gen 1)

B

Far end of *EPIDIDYMUS* (distal)

B **UPDATED MiRNA in Sperm**

During the passage from **A** to **B** "...multiple waves of small RNAs (miRNAs and tRNAfs,
etc.) are synthesized or loaded into sperm..." By the far end (distal) of the epididymus
the gains include: miRNA-34b/c, miRNA-15, 16, miRNA-17-92 cluster, miRNA-880
cluster and tRNAf-Val-AAC, tRNAf-Val-CAC.

Sharma et al., "Small RNAs Are Trafficked from the Epididymus to Developing
Mammalian Sperm" (2018) *Developmental Cell* **46** 481-494.[5]

"Using the TU tracer system we show that miRNAs first synthesized in the
epididymus **A** make their way to maturing sperm, **A** to **B** thus
demonstrating that **soma-to-germline** transfer of small RNAs [MiRNAs,
tRNAfs, etc.] occurs in intact animals."

Figure 45: MALE LEGACY: Male miRNA

BRIEF: 46

Across the top of the illustration in *Figure 46*, the stages of cellular division of the embryo are portrayed in the journey from zygote to implantation. Hassam Khatib guided his University of Wisconsin team of Nicole Gross et al., in the development of "Micro RNA Signaling in Embryo Development," *BIOLOGY* (2017).[6] An insightful picture of miRNAs in communication (cross-talk) between the embryo and mother is provided. During this 6-to-12 day stage, the embryo's traverse through the liquid information-rich environment of the fallopian tube is a storm of information exchanges inside and surrounding the embryo.

In explaining the transition of miRNA from its first form, Pri-MRA, to the mature miRNA's ability to express genes and the structure of proteins, the Khatib-Gross team is, in my opinion, pointing to a powerful potential. This male and female navy of small miRNAs may have the power to *reconcile the unrelated ancestral parental genes* and integrate the two unique streams of adaptive information they possess.

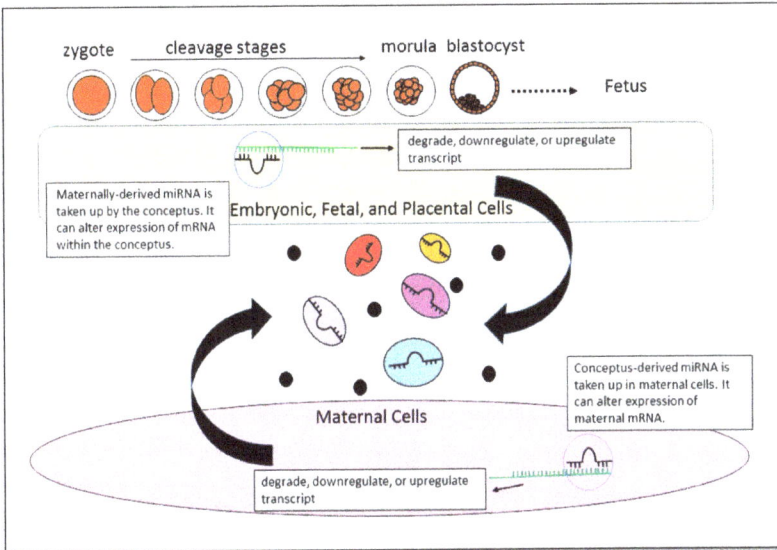

The Firestorm of Female miRNA Extracellular Expression—Zygote to Implantation.

Nicole Gross et al., "Micro RNA Signaling in Embryo Development" *Biology* 2017 **6**, 34[6]

"…the Zygote has the ability to differentiate into any cell type in the body. To facilitate this cellular transition, extensive molecular remodeling must occur during fertilization and early embryo development (zygote to blastocyst)."

"Expression of microRNA (miRNAs) is essential for embryonic development… miRNA's are secreted in the extracellular environment {biological fluids} by the embryo during the preimplantation stage of development."

Gross, et *al.*, focused on the role of miRNA's in "signaling in the context of preimplantation embryonic development and embryo-mother cross-talk, with an emphasis on modes of transfer for miRNA's."

Figure 46: FEMALE LEGACY: Female miRNA

BRIEF: 47

A working assumption consistent with the
Theory of Co·GENESIS

The acknowledged weakness of the DNA-centered model is the premise that the random joining of egg and sperm upon fertilization accounts for the variability and adaptation observed across species. Many scientists agree that this is a mathematical impossibility. The strength of Co·GENESIS is in the exponentially greater variability through the two lines of parental environmental adaptation and the randomized synthesis of those lines across the hemispheres of the brain. Darwin concluded that such high levels of variability must be present in order for Natural Selection to achieve the complexity and speed of evolution he observed.

Rather than a single merger of egg and sperm, two Stages are proposed. In Stage One, RNA reconciles the two parental codes as the RNA Body prior to expression of the Applied RNA MIND in Stage Two. The Stage One window of access to the separated parental genomes (2-cell to 16-cell) allows male and female miRNAs to update the two parental lines before they are closed. Using illustrations from "Gray's Anatomy" as a base, *Figure 47* describes Stage One. Parental miRNAs work within the membrane to reconcile the Codes, and as shown in the 32-cell, *establish the new male/female binary cells*. During Stage Two, after the closure of the 16-cell, long non-coding RNAs (lncRNAs) begin to connect imprinted control regions (ICRs) of the Body[7] and synthesize a new Code for the neurological and endocrine systems—the applied RNA Mind.

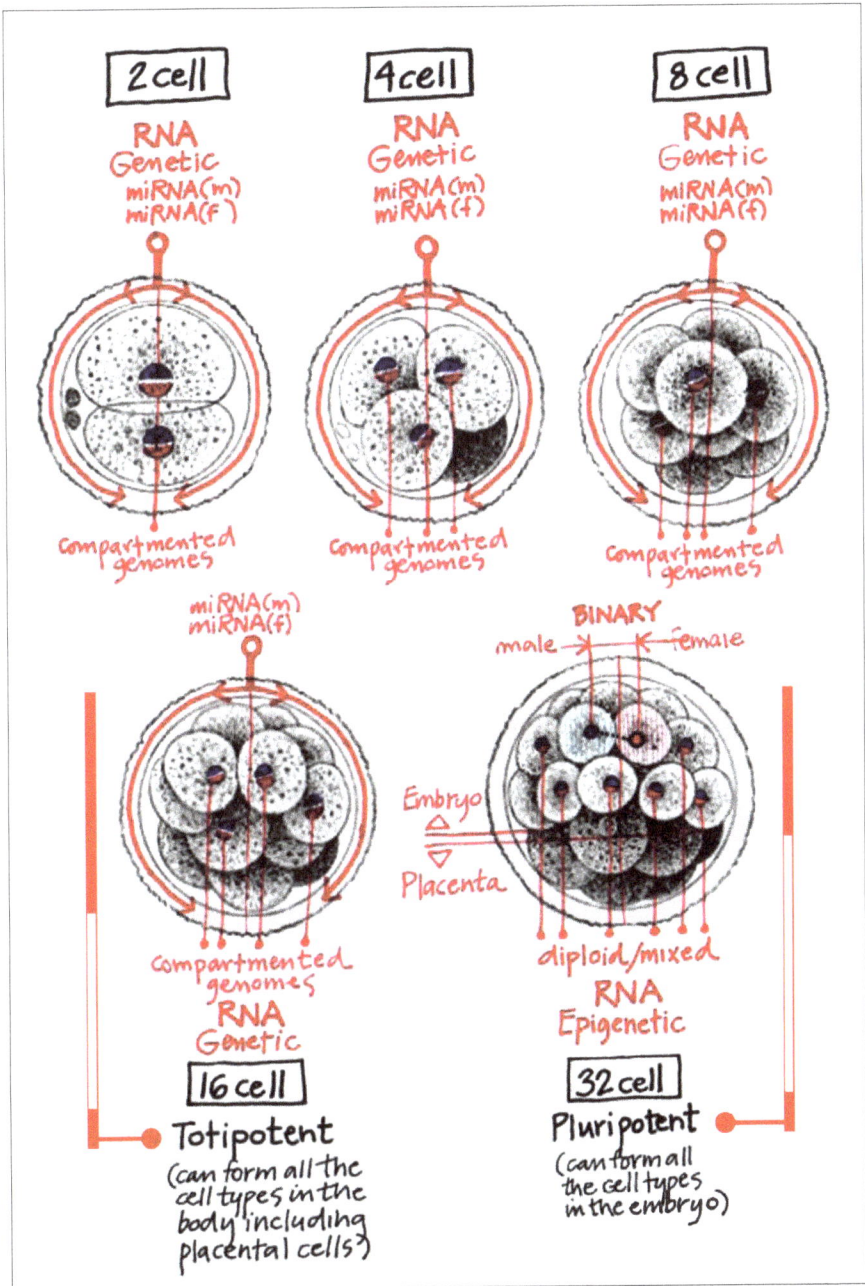

Figure 47: RNA RECONCILIATION of the 2-cell to 32-Cell Embryo.

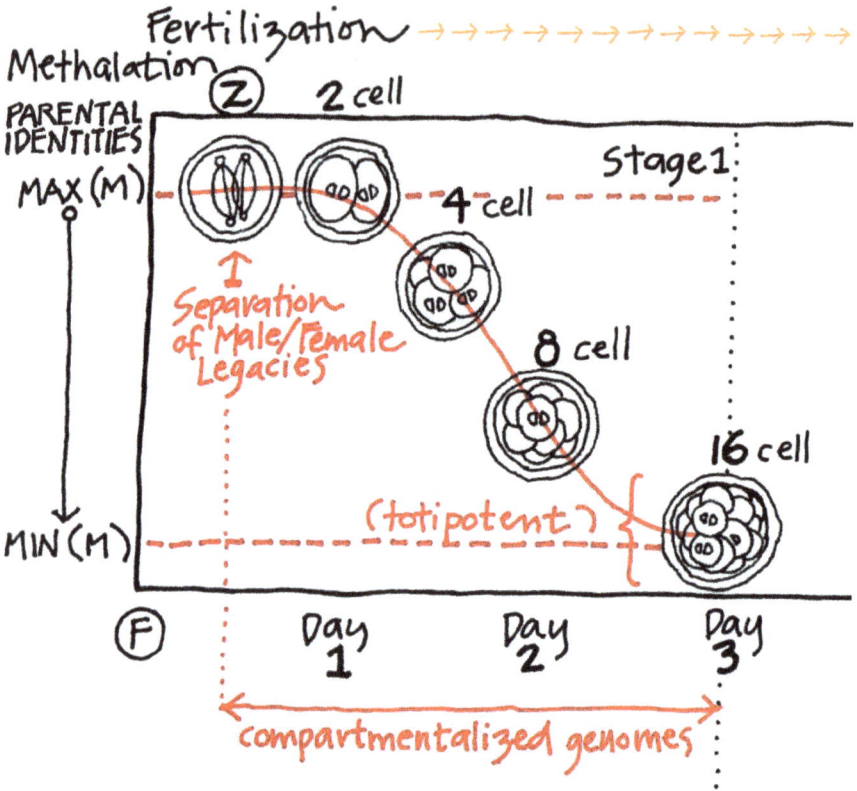

Figure 48. STAGE ONE Genetic/Body Demethalation: Removing Parental Identities

BRIEF: 48

A working assumption consistent with the Theory of Co·GENESIS.

STAGE One: The zygote contains the genomes of two unrelated individuals (incest avoidance); their separate adaptive information and lineages must be reconciled and recast for the Code of Life of a single individual (offspring). Information within the genomes is primarily reconciled via miRNAs and the parental identities (methalations) are removed while the genomes are "open." At the 16-cell division, the updated RNA Body cells will be mixed (diploid) marking the end of Stage One. At this point the new male/female cells of the Binary are also established and Stage Two, casting a new Identity begins.

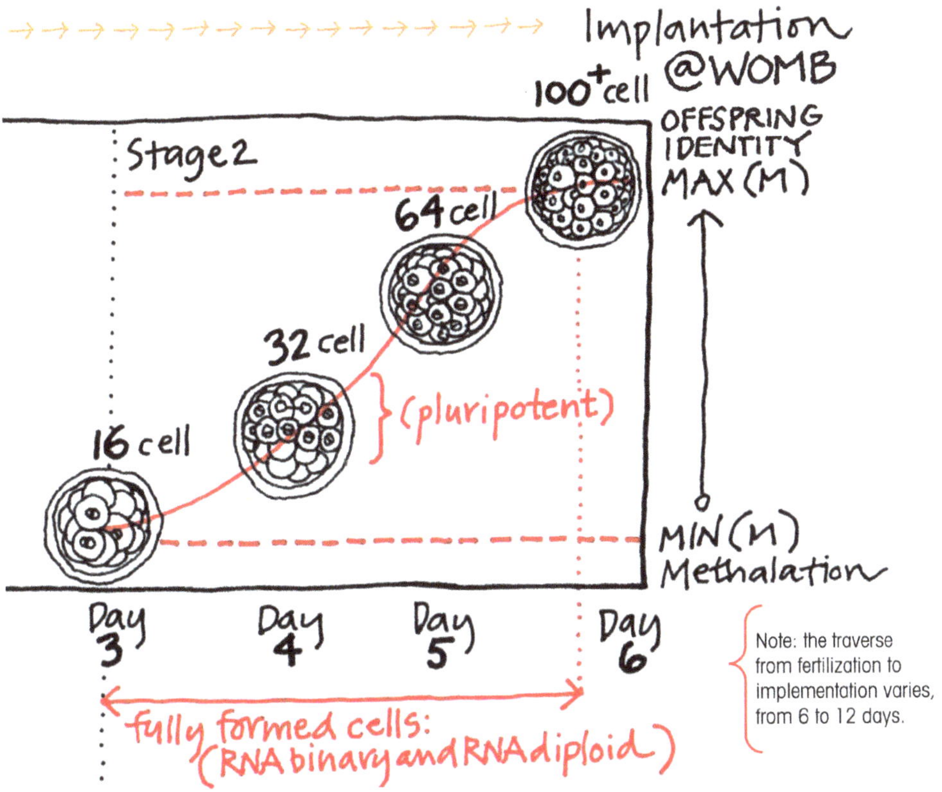

Figure 49. STAGE TWO Epigentic/Mind: Remethalation, a New Offspring Identity

BRIEF: 49

A working assumption consistent with the Theory of Co·GENESIS.

STAGE Two: Male and female miRNAs were dominant in Stage One when the mission was to merge the two parental Codes for an RNA Body however, long non-coding RNAs (lncRNAs) become key to Stage Two, where the mission is to establish a new methalation/identity for offspring. The lncRNAs establish imprinting control regions (ICRs): clusters of 2 to 20 imprinted genes primarily in the brain but interconnected across the Body as a master regulatory system. Following Stage Two, the embryo is implanted in the womb and undergoes ±4 months of gestation before the RNA MIND has progressed adequately for RNA-genomic imprinting. Imprinting passes male updates to the (baby) sperm or female updates to the (baby) eggs—confirming that a basic transgenerational RNA MIND exists at ±4 months.

BRIEF: 50

The RNA MIND is who we are and it exists primarily on two interconnected axes: the first is the HPG Axis linking the hypothalamus and the pituitary gland and gonads (the gametes, etc.) which can be thought of as the *reproduction domain*. The second axis is the HPA Axis linking the hypothalamus, the pituitary gland and adrenal gland which can be thought of as the *survival domain* (not shown here). *Figure 50* is the male HPG Axis, generally described by Maruska and Fernald in their paper for *PHYSIOLOGY* (2011).[8]

The HPG and HPA Axes are closely related in the sequence of their embryonic formation (as the *adrenal-gonad primordium*) and they are both coordinated through the hypothalamus—the "brain within the brain."

Notably, in Chapter 4, the team of Sankararaman et al. identified Neanderthal (HN) miRNAs (messengers within the HPG & HPA Axes) as being "depleted," and the HN adrenal gland of the HPA Axis as being the second lowest of sixteen tissues tested for commonality with Humans. Further, they describe RNA processing—high levels of which are required for Human deliberative consciousness—as "depleted" in Neanderthals. This chapter has provided convincing evidence at the molecular level supporting the cultural and archeological evidence in Chapter 4. Deliberative-consciousness, present in Humans, was most unlikely to be present in Neanderthals when the two apex predators faced each other in Eurasia-Western Europe ±50-42,000 YBP.

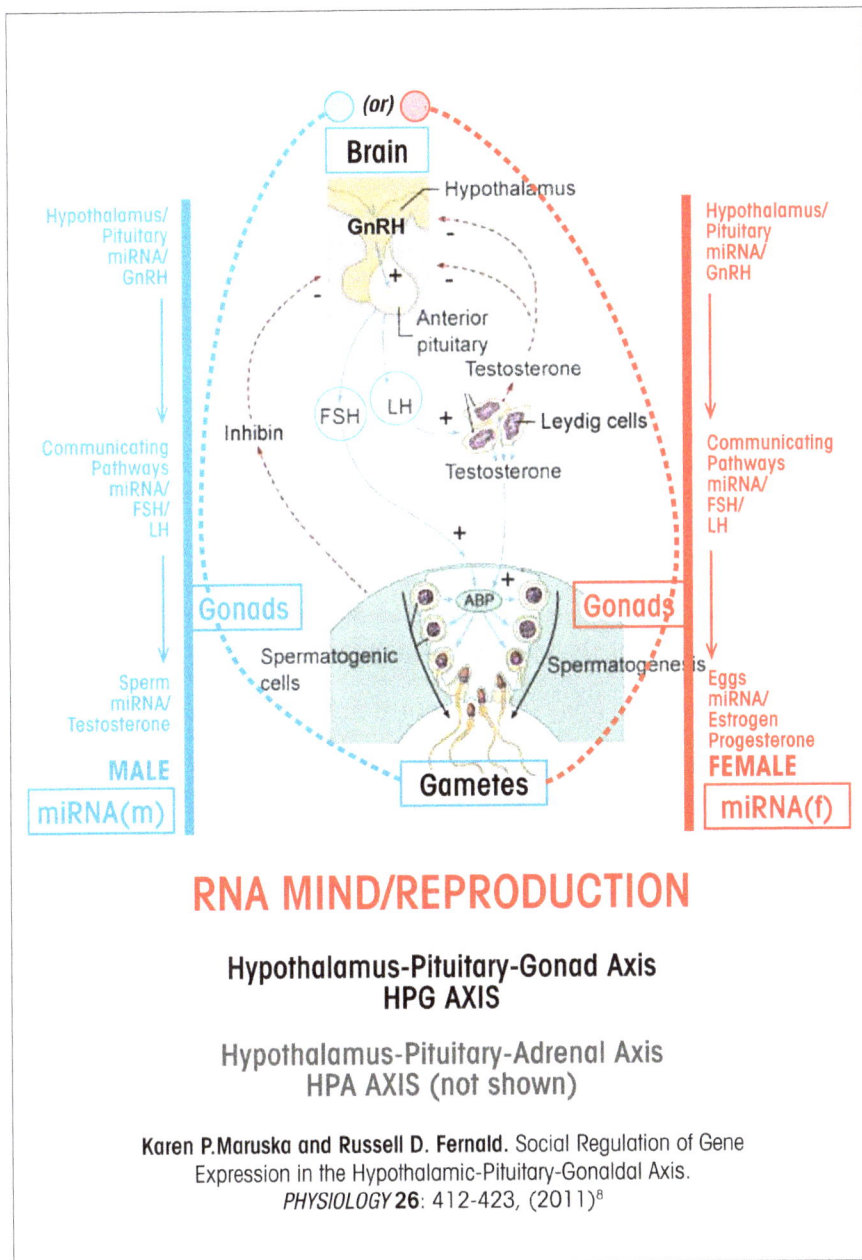

Figure 50: RNA MIND/REPRODUCTION: Male and Female Pathways

BRIEF: 51

Look beyond the complexity of what is going on in *Figure 51*; the seven red circles that enclose the miRNA messenger molecules tell the whole story. Essentially the RNA MIND acting through these miRNAs, is expressing the universal Body as male or female. Cao et al. used this illustration of a gonadotrophic cell in their article in *Molecular and Cellular Neuroscience* (2018)[9] to argue that miRNAs are "undervalued" given their defining roles in reproduction and the neurological system/endocrine system.

In this example, ten small RNAs initiate the expression of the BODY as Male or Female (HPG Axis). Through participation in the life of their male or female host, [miRNA(f) or miRNA(m)] are uniquely positioned to pass down environmental adaptations and reconcile the male and female genomes in Co·GENESIS. The Cao team also highlights the uniqueness of miRNAs from one species to the next. *Figure 51* provides an excellent final image to summarize Chapter Five:

RNA is the verb (acting)
DNA is the noun (acted upon).

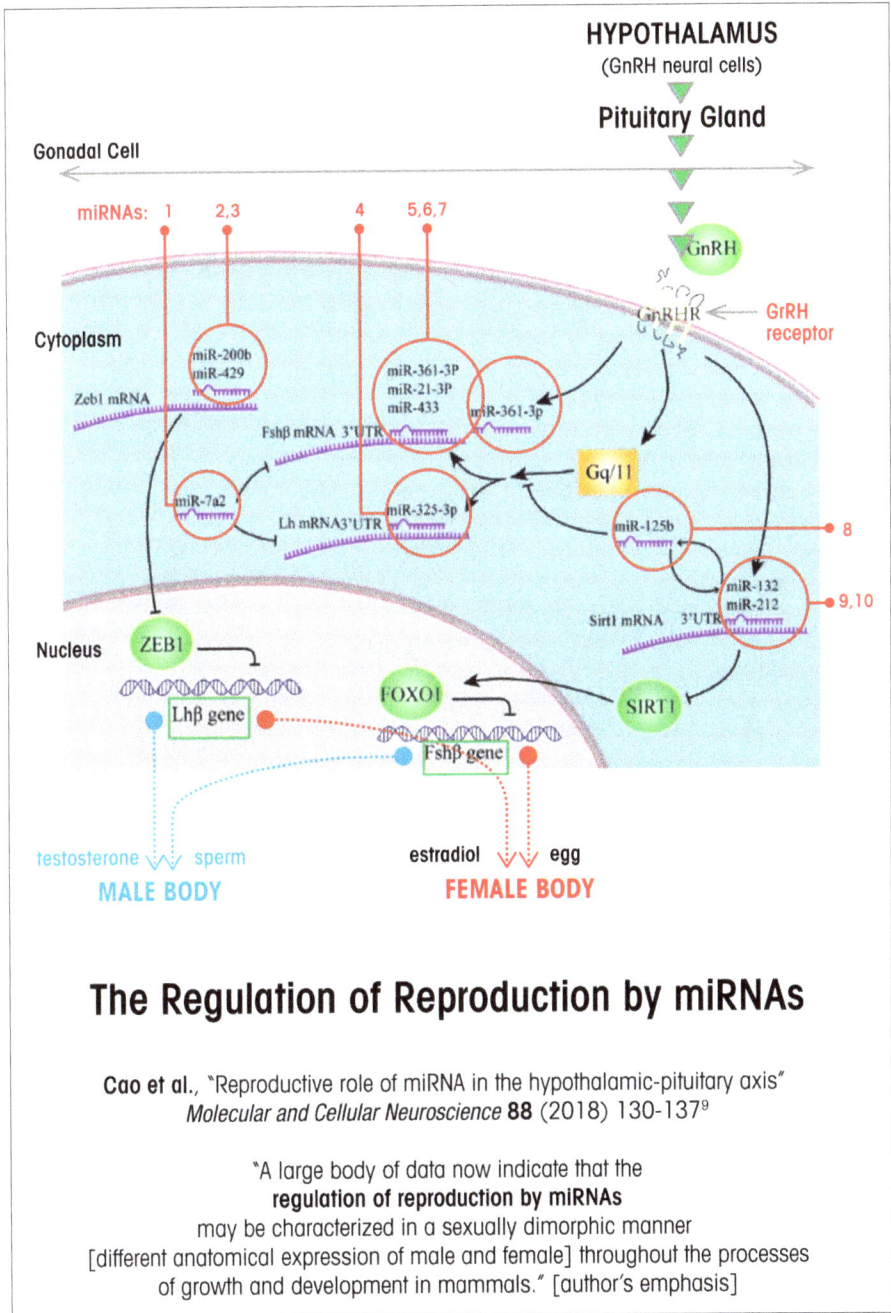

The Regulation of Reproduction by miRNAs

Cao et al., "Reproductive role of miRNA in the hypothalamic-pituitary axis" *Molecular and Cellular Neuroscience* **88** (2018) 130-137[9]

"A large body of data now indicate that the
regulation of reproduction by miRNAs
may be characterized in a sexually dimorphic manner
[different anatomical expression of male and female] throughout the processes
of growth and development in mammals." [author's emphasis]

Figure 51: RNA Initiates Male or Female Interpretation of the Body

CHAPTER SIX

TRANSGENERATIONAL PATHWAYS

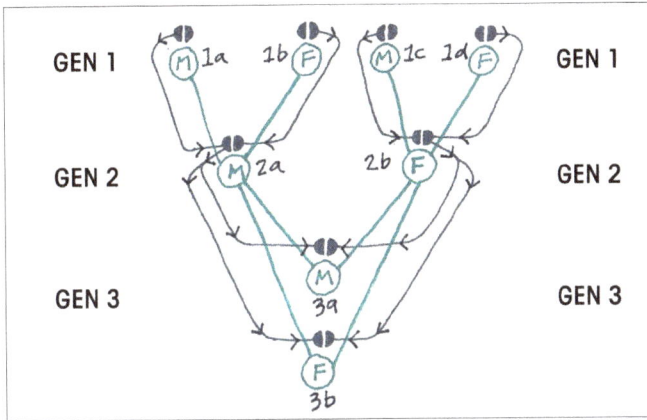

Figure 52. RNA Epigenetic/Genetic

We have examined the mechanisms of perception, behavior, and consciousness that are consistent with an RNA-centered world. However, in this Chapter we are going to seek a multi-generational sense of how we experience the RNA MIND passed down through us—an operational sense of being Human. We begin with the three-generational diagram (See *Figure 52*) from the Foreword. Drawing on the work of seven international research teams, we flesh-out the diagram to see how we are continuously responding to, and being shaped by, our environment. We conclude with the teams of Maatsura (2020), and Inneweber (2020), whose research papers, I would argue, provide third-party support, if not validation, of Co·GENESIS.

The RNA MIND of a species is difficult to imagine because, until fertilization, it exits as two unrelated halves of a future MIND being prepared over and over in the male and female of the species as gametes in anticipation of conception. Each gamete is constantly receiving information along the way, information that may one day update the ancestral male and female legacy in a future union. Prior to conception there is a near-infinite potential for variation. However, after fertilization, every time one of the variables, even a single nucleotide, falls into place, one direct possibility has been elected and many other indirect possibilities have been eliminated. In this manner, there is a paring down during the reconciliation between the two parental lines of ancestral and adaptive information resulting in unique offspring. Randomized individual hemispheric interconnections are achieved and yet a resilient universal pattern, the social order, is maintained across the species as a whole.

BRIEF: 53

Figure 53 is a blow-up of *Figure 52*, showing the flow of adaptive information from the environment as it passes through three generations. The environments, unique life experiences in each generation, are called out on the left as ⓔ, ⓕ, and ⓖ, and the information flow via gametes across generations is shown as male (blue) and female (red). For instance, the couple in the upper left (D_1, D_2) had their Applied RNA MINDS updated by their parents who lived in environment ⓓ and their son (E_1) will be updated by his parent's experiences in environment ⓔ. The gamete (sperm) of E1 will update the left hemispheres of his son and daughter based on his experiences in environment ⓕ.

Adaptations passed down can be thought of as provisional due to the process of *assimilation*. If the environmental condition that gave rise to an epigenetic adaptive change continues in subsequent generations, it becomes a genetic modification; if not, it fades away. Ken Nishikawa and Akira R. Kinjo, writing in *Biophysical Reviews* in 2018,[1] referred to this adaptive process as the *plasticity* of an individual's characteristics in response to environment. Jean-Baptiste Lamarck and Charles Darwin predicted this gain or loss of adaptation due to the continuing interaction of organisms within their environments.

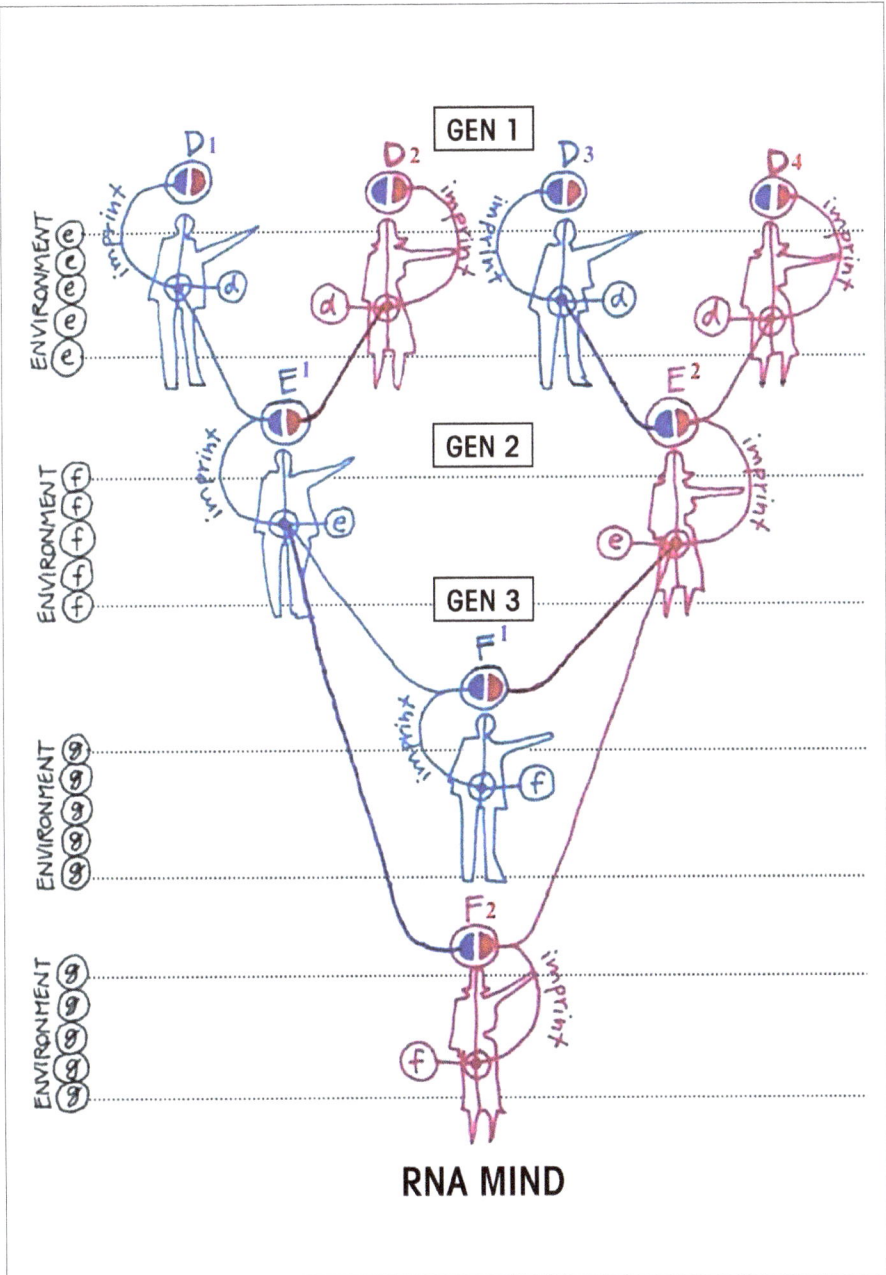

Figure 53. ENVIRONMENTAL ADAPTATION via RNA Imprinted Gene Flow

BRIEF: 54

Co·GENESIS appears to be applicable to all cellular, nucleated life that reproduce sexually. Therefore, *Figure 54* is intended to be broadly applicable, showing how the male and female legacies of species maintain a male-to-male M1-M3-M4 (blue) and female-to-female F2-F3-F4 (red) transgenerational continuity. The end of continuity is also shown; when a parent has only offspring of the opposite sex (F1 to M3) and (M2 to F3), there is no continuity to the GEN 3 offspring. The biological intervals of life shown here are from birth of parent to the giving of birth by the parent (generally 20 to 40+ years in humans), the durations of adaptive input vary (e.g., 22 years for M4, 40+ years for F4).

The Legend is a close-up of the second-born in the third generation, describing the *Applied RNA MIND*, the *Anticipatory RNA MIND*, and the interactive RNA neuroendocrine process of updating the gametes. Life is constantly becoming, and the Anticipatory Mind is where life experience (adaptation) is transferred into the future.

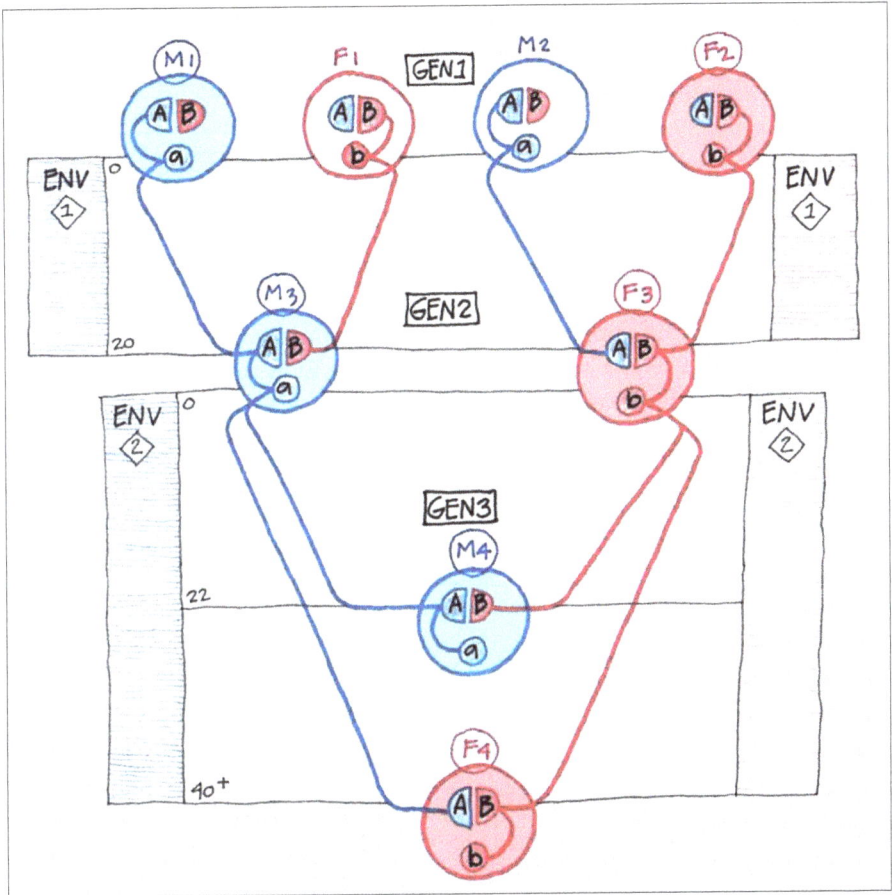

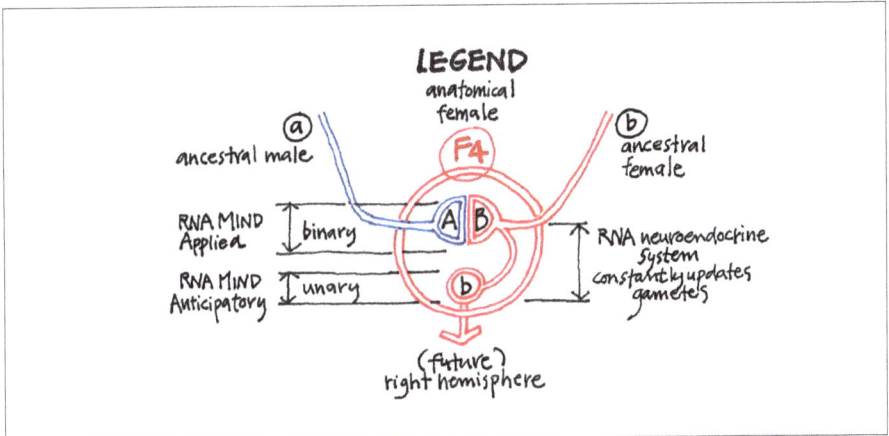

Figure 54. THE RNA MIND: Applied, Anticipatory, and Pathways

BRIEF: 55

THE ORDER OF LIFE

1. BODY: The RNA body simply evolves through Natural Selection toward the best physical form to meet the challenges and opportunities in the surrounding environment (heritable/unary). For instance, night vision is one such trait, which we reviewed earlier; it is shown as light green in *Figure 55/Body*. (Convergent process).
2. MIND: The RNA perception-behavior of individuals in a species evolves to an overall resilient pattern of variation, to best meet challenge and opportunity in an indeterminate future (non-heritable/binary): *Figure 55/Mind*. (Divergent process).

Survival is the first objective of a species and there are extinction-level events that must be addressed in any species if it is to survive over time. Humans face the possibility of a meteor strike, or nuclear exchange, or mega-volcano events, any of which can create years-long nuclear winter in which mass starvation triggers a die-off of ±90% of the population and a life-or-death struggle for resources. A civilized remnant population that would rather die than kill others in the fight for food would go extinct. The presence of males and females who can kill without remorse and eat human flesh before starving, could pull a species through this darkest hour. And most importantly because of Co·GENESIS, those *persistent survivors* (males and females) will bear offspring, including the humanists, the inventors, the writers, the poets, and the musicians of the ensuing population. The social order's spectrum of perception/behavior is built-in and can be restored within 5 generations vs. 1,000 generations for changes to the RNA Body.

A resilient species in crisis, like a sailboat, leans far over in the storm, spilling air to remain afloat, and then rights itself as the storm recedes. Such resilience and diversification means just a few astronaut couples could populate an earth-like planet, and eventually produce another 7.8 billion unique Minds.

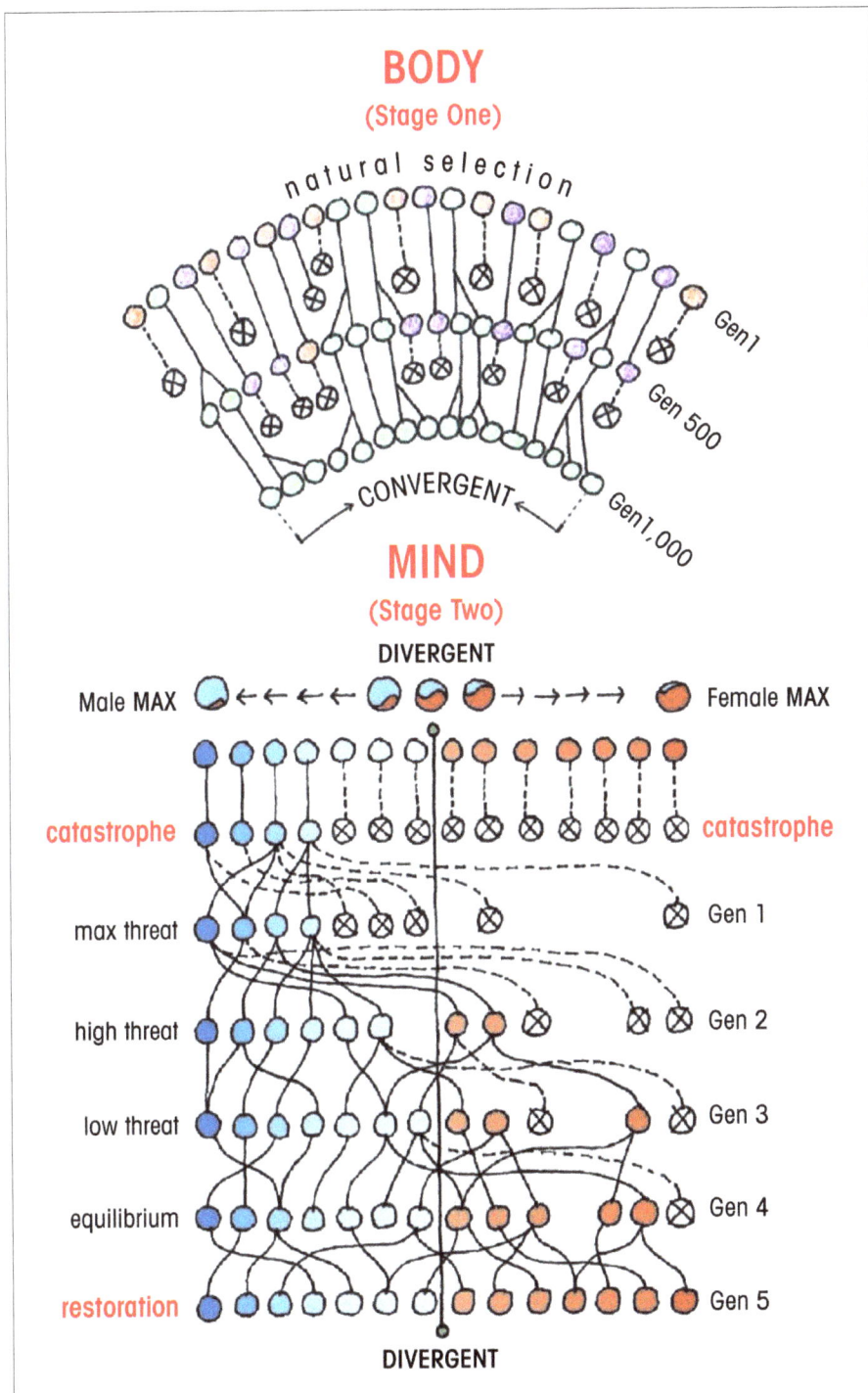

Figure 55. THE ORDER: Body/Heritable & Mind/Not Heritable

BRIEF: 56

An overview of male and female imprinted genes, with distribution across MIND and BODY is shown in *Figure 56*. Olivia Ho-Shing and Catherine Dulac with Harvard's Center for Brain Science adapted this image from the earlier version by Perez et al. (2016) for their article on brain function and behavior, published in 2019.[2] On the right are 28 imprinted genes, and moving from right to left we see the variable expression of each gene in 7 parts of the body and 16 regions of the brain. The blue shades are male-legacy biased, ranging in intensity from pale blue to dark blue; similarly, the red shades are female-legacy biased, ranging from pale red to dark red. The take-away is that our MIND is a two-part grouping of imprinted genes: a small, highly intense, *maleness domain* and a much larger, but less intense, *femaleness domain*. The BODY is further to the right, without the compelling male-female signature.

Quotes from HO-SHING and DULAC:

1) "…consensus has emerged that imprinted effects are more prevalent in the brain than somatic [Body] tissues in mice, rats and humans, supporting the notion that genomic imprinting may constitute a conserved mechanism to instruct neural [Mind] function."

2) "Multiple studies have characterized higher frequency of imprinted gene expression in the hypothalamus (Hy) than in other brain regions…"

3) "Many imprinted genes play further roles in the adult brain by regulating synaptic transmission and plasticity [forming and retrieving memories]."

Ho-Shing and Dulac concluded: "These genes form extensive regulatory networks that influence the physiology [functioning] of neural circuits [the Mind] and affect behavioral phenotypes [personality] in the adult."

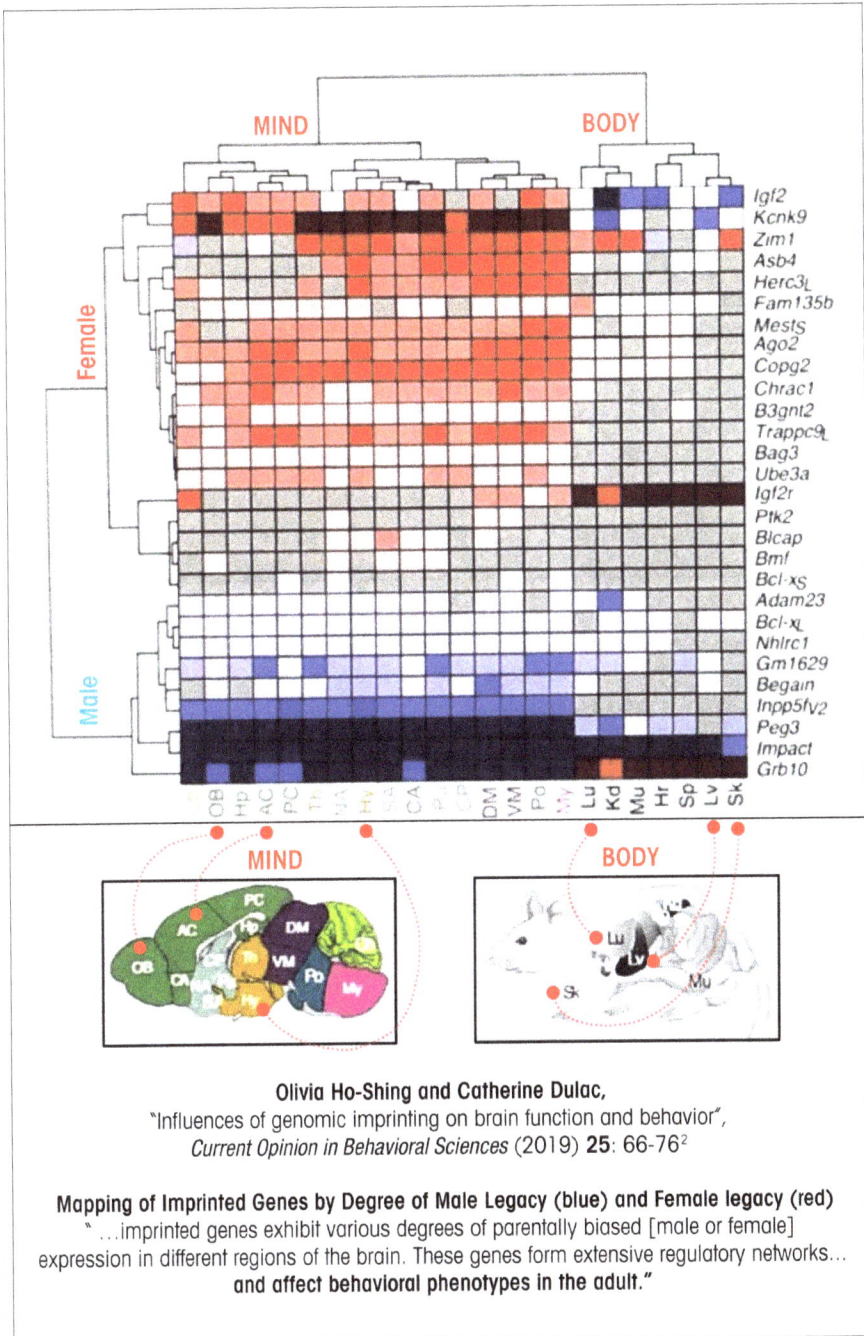

Olivia Ho-Shing and Catherine Dulac,
"Influences of genomic imprinting on brain function and behavior",
Current Opinion in Behavioral Sciences (2019) **25**: 66-76[2]

Mapping of Imprinted Genes by Degree of Male Legacy (blue) and Female legacy (red)
" ...imprinted genes exhibit various degrees of parentally biased [male or female]
expression in different regions of the brain. These genes form extensive regulatory networks...
and affect behavioral phenotypes in the adult."

Figure 56. 28 Imprinted Male and Female Genes of Mind and Body

BRIEF: 57

THE MASTER

The hypothalamus, the center of the Applied RNA MIND, is illustrated in Figure 57, located on the brain's centerline with its 28 imprinted genes from Olivia Ho-Shing and Catherine Dulac's matrix displayed below. There are the 9 male legacy genes (blue) which are shown to the left of the centerline and the 15 female legacy genes (red) to the right. This reflects the distribution of their influence, but not necessarily their anatomical position in relation to the brain because they can exert influence at great distances. The numerical values (reflecting the strength of parental bias/expression) are shown in the bar graph below the centerline. It appears that males and females with a strong left hemisphere expression (maleness-blue) tend to assertively perceive a simpler black-and-white world while males and females with a strong right hemisphere expression (femaleness-red) tend to perceive a more complex world with gray tones and less assertiveness.

The hypothalamus conducts this binary orchestra, expressing the RNA MIND through the regulatory networks of functional RNAs, including lncRNAs, imprinted genes, and miRNAs, plus electrical (neural) and hormonal (endocrine) signaling. Early on, researchers, seeking to understand the two hemispheric roles induced the joining of mouse embryos that do not occur in nature. One was formed from two half-sets of the male-legacy and one was formed from two half-sets of female-legacy. Neither was viable, but researchers could see that the all-male embryo was developing a large body with a small brain, and the all-female embryo a small body with a large brain. The explanation for these variations of the body and the hemispheres of the brain would have to wait (in my opinion) for a reconceptualization: Co·GENESIS.

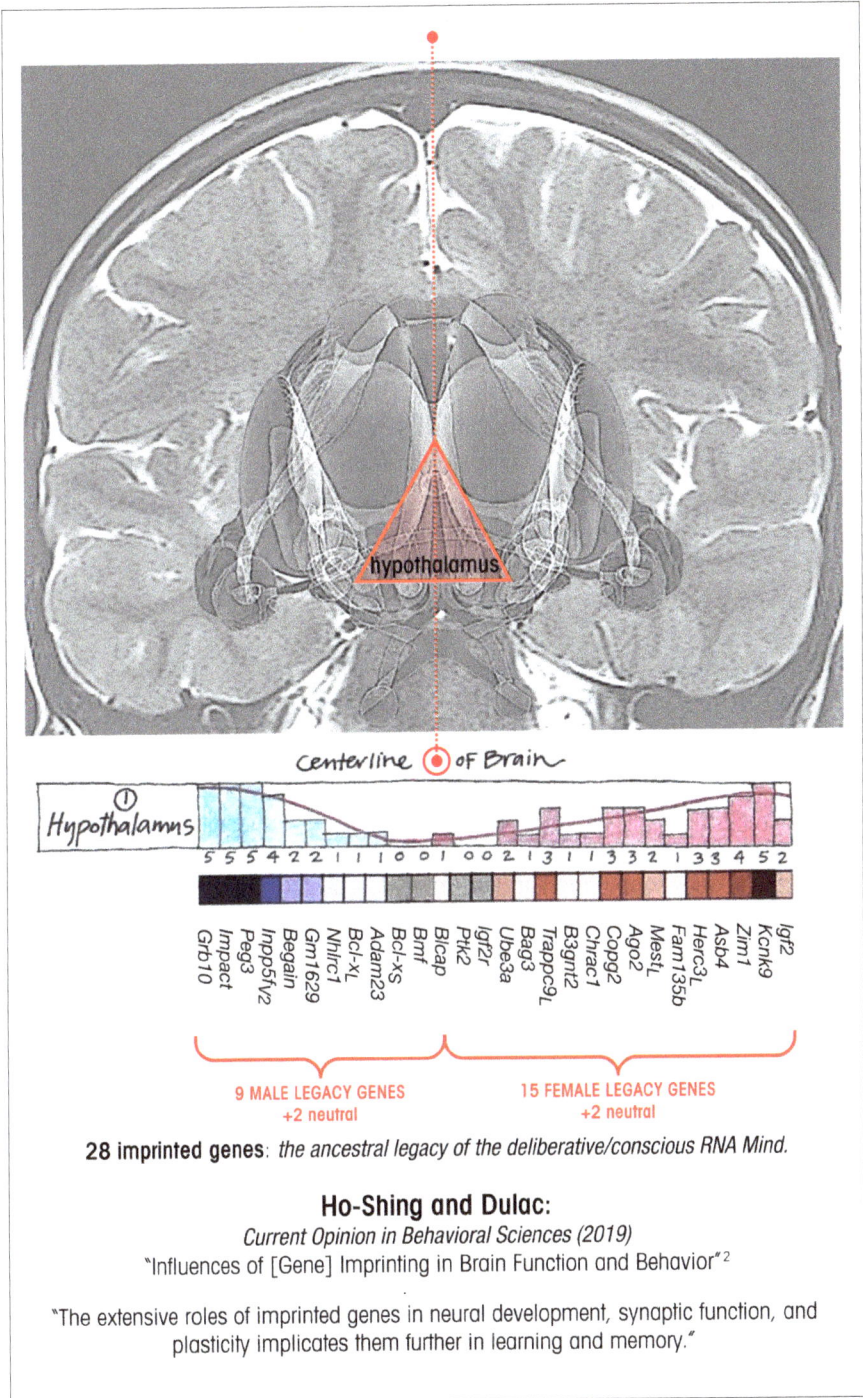

Ho-Shing and Dulac:
Current Opinion in Behavioral Sciences (2019)
"Influences of [Gene] Imprinting in Brain Function and Behavior"[2]

"The extensive roles of imprinted genes in neural development, synaptic function, and plasticity implicates them further in learning and memory."

Figure 57. HYPOTHALAMUS: Master of Survival & Reproduction

BRIEF: 58

CONSCIOUSNESS

In Chapter 4, on page 54, I linked the binary origin of consciousness to what I believe to be the under-appreciated work of the Max Planck Institute for Human Cognitive and Brain Sciences research team of Daniela Sammler et al., published in *Cell* (2015).[3] See the upper illustration of the hemispheres of the brain in *Figure 58*, which is my conceptual mirroring (left-right) of their image to represent the right hemisphere pathways of speech, paralleling those of the left. Up until their research (speech typically being attributed solely to the left hemisphere), the argument for speech/language as the biological bridge of deliberative consciousness was not fully supported.

By utilizing brain imaging and functional tracking (fMRI), the team mapped brain activity in real time associated with the auditory response to two words: "bear" and "pear." The appeal of the two words is that they sound alike, requiring the brain to distinguish between the sounds for the meaning of the words. Additionally, the subjects were asked to speak the words as a fact, and then as a question, which requires vocal tone (emotion and intent). This latter melodic variation is called *prosody*, critical to understanding and social interaction. The researchers confirmed "that prosody perception takes dual routes along dorsal and ventral pathways in *the right hemisphere.*" The pathways of vocabulary, speech, and meaning, exclusive to the left hemisphere, therefore have essential parallel pathways in the right hemisphere. The *maleness* of the logical left hemisphere pathway opposite the *femaleness* of the emotional right hemisphere pathway imparting meaning is an exemplar of the binary bridge of Co·GENESIS.

The Sammler team used the most fluid and universal of our human traits—perception of language and meaning—to isolate views into that collaboration. The RNA MIND's neurological architecture and intertwining of these two hemispheric world-views are further evidence for the Human evolution to cross-talk, (deliberative consciousness), and the co-evolution between and among humans that followed.

The Nature of Consciousness

a binary biological intelligence

(Note[1]: the fMRI neuro-images in the hemispheres below are "mirrored" to represent functional parallel tracks)

Female Legacy Parallel   Male Legacy Parallel

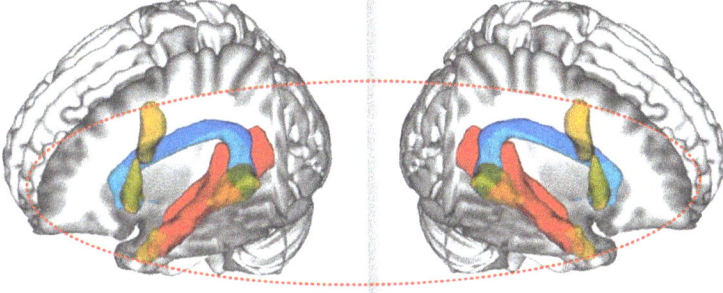

Right hemisphere
auditory cortex

**Frontal
View**

Left hemisphere
auditory cortex

Distinguishing vocal tone: emotion and intention
(by rhythmic and melodic variation: *prosody*)

Understanding ... "the speaker's emotions
and intentions— thereby making
it an important tool in social interaction."

Distinguishing one word from another
(by the unit of sound: *phoneme*)

Understanding the arrangement of words and
phrases (syntax) and the
logical aspects of meaning (semantics).

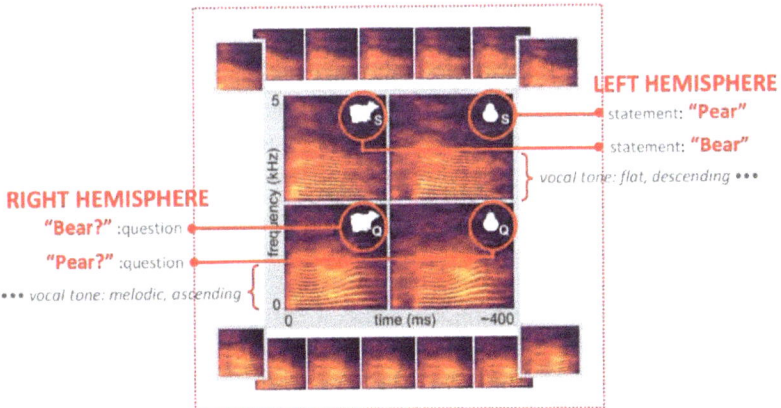

Daniela Sammler et al.,
"Dorsal and Ventral Pathways for Prosody"
Cell **25** Issue 23 (December 2015) p. 3079- 3085 doi: 10.1016/j.cub.2015.10.009[3]

Figure 58. The Nature of CONSCIOUSNESS

BRIEF: 59

The cyanobacterium, its RNA in sync with circadian rhythm, slowly produced the oxygen-laced atmosphere that supports complex life on Earth. Tonight, you will subconsciously carry out another RNA daily process by becoming more precisely you, more aware of what is most important to you. The phenomenon of REM sleep is illustrated in *Figure 59*, based on the work of the Elissa Pastuzyn Team (2018)[4] and consistent with the Izawa (2019)[5] teams. During REM, the hypothalamus and its MCH neurons "weed the garden," taking away the insignificant by forgetting and resetting the synapses for new information. Significant short-term memories, made out of mRNA, are packed into extracellular vesicles of viral origin, to float away into long-term memory. This is the domain of the RNA MIND about which scientists know little, and which has only recently emerged from the title "Junk," to the category of Dark Matter. Our state of affairs would not be so concerning if a scientist, He Jiankui, had not already introduced multiple corrupted genes (CCR5) into the Human germline, genes with functional roles in the RNA MIND.

Who are we if not our memories, the collective expression of our pathway through life? We cherish memories that we relive with family and friends, and find no greater cruelty in life than to have the mind of someone we love taken away. Anyone who has lost a loved one to Alzheimer's is a witness to the slow reversal of all that their RNA MIND curated...the undoing of who they were. Conversely, we marvel at the 12-to-18-month-old child already expressing the complex language of their birthplace—not learning, but remembering, as Plato said in reference to the Immoral Mind—the Mind that comes down to us after a billion years in the making. Chapter 7, "Significance and Meaning," will seek to frame the imminent threat to our consciousness and, therefore, our viability as a species.

In the manner of Reichmann et al., three additional research teams have published important parallels, if not validations, of Co·GENESIS. I submit these for your consideration as VALIDATION BRIEFS 60 (2017) and 61, 62 (2020) which conclude this Chapter.

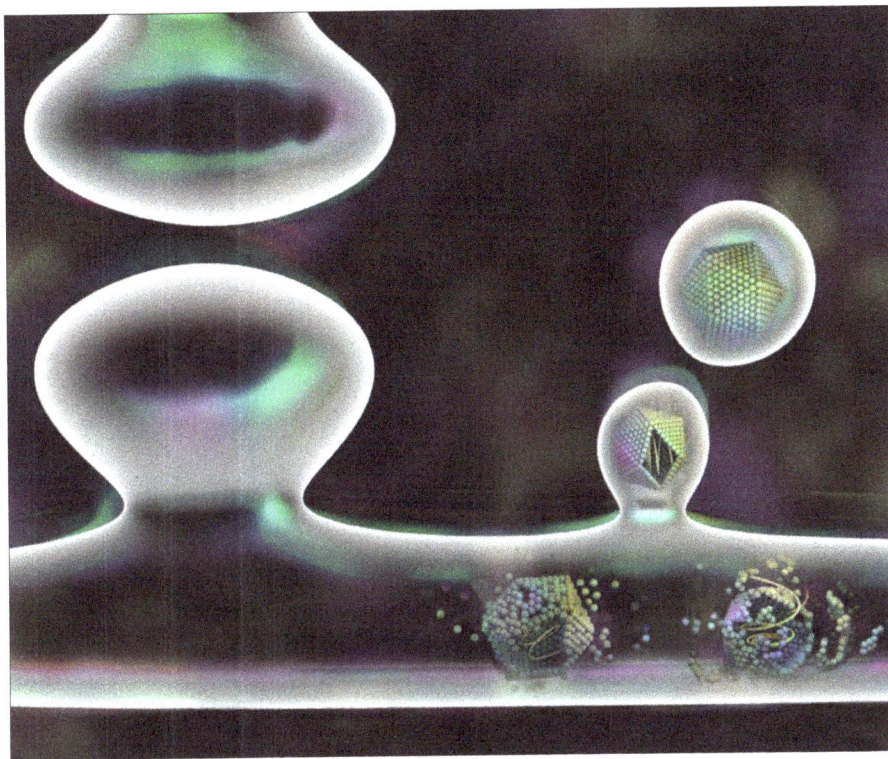

"Virus-like shells budding off from one neuron and moving to another."

Elissa D. Pastuzyn et al.
"The Neuronal Gene Arc Encodes a Repurposed Retrotransposon Gag
Protein that Mediates Intercellular RNA Transfer" *Cell* **172**, 275-288 (2018)[4]

"Long-term information storage in the mammalian brain...*Arc* protein is released
from neurons in *extracellular* vesicles that mediate the transfer of mRNA
[short-to-long-term memory] into new target cells."

Memory is edited by the RNA Hypothalamus during REM Sleep

S. Izawa et al.
"REM sleep-active MCH neurons are involved in forgetting (and acquiring) memories."
Science **365**, 1308-1313 September 2019[5]

"Events experienced during wakefulness are stored as memory...these memories
undergo selection (editing) during sleep...MCH neurons in the **hypothalamus**
actively contribute to forgetting in rapid eye movement (REM) sleep."

Figure 59. RNA NIGHTLY: Consolidation and Editing of Memory

VALIDATION BRIEF: 60

"This is an important challenge for evolutionary theorists."

With these words, David Houle et al., writing on the evolution of the wing patterns of Drosophilids (fruit flies), had my undivided attention.[6] See *Figure 60*. After examining over 50,000 fly wings and the high degree of variation around 12 landmark vein intersections, the team found that, adjusted for size variation across species, all fly wings are similar even after 40 million years of evolution. A high degree of variation with such a low rate of evolution was "not readily explained by available models of evolution."

The Theory of Co·GENESIS predicts just such a pattern for well-adapted organisms in stable environments over long periods. Each individual fly, male or female, lives a unique life experience generating epigenetic adaptation and variation in offspring. Due to genetic assimilation, such changes will become more permanent or drop away in subsequent generations if they do not provide an adaptive advantage. Organisms in stable environments, or ones for which they are well-adapted, (such as the shark and the fruit fly) will exhibit a pattern of "spontaneous mutational variation" in each generation. But over time, if these prove to be unnecessary, they will fall away, resulting in little or no long-term evolutionary change.

(A) Variations in wing form
@ landmark intersections:
50,000 wings analyzed.

Legend:
Immigrans-tripunctata △
Zapnonus △
Hirtodrosophila △
Mycodrosophila △
virilis-repleta △
Scaptomyza △
Idiomyia ▽ (silvestris ▽)
Sophophora ● (melanogaster ○)
Chymomyza □
Scaptodrosophila □
Steganinae □

Houle et al., "Mutation predicts 40 million years of fly wing evolution"
Nature **548**, 447-450 (2017) doi.org/10.1038/nature23473[6]

M_{hom}
M_{het}
G

(B) Ellipses mark patterns of ***spontaneous mutational variation***
occurring within and across the fruit fly varieties (see above).

*"Our finding of striking similarities among mutational, genetic and among-species
variation coupled with a low rate of evolution are not readily explained
by any of the available models of evolution.*
This is an important challenge for evolutionary theorists."

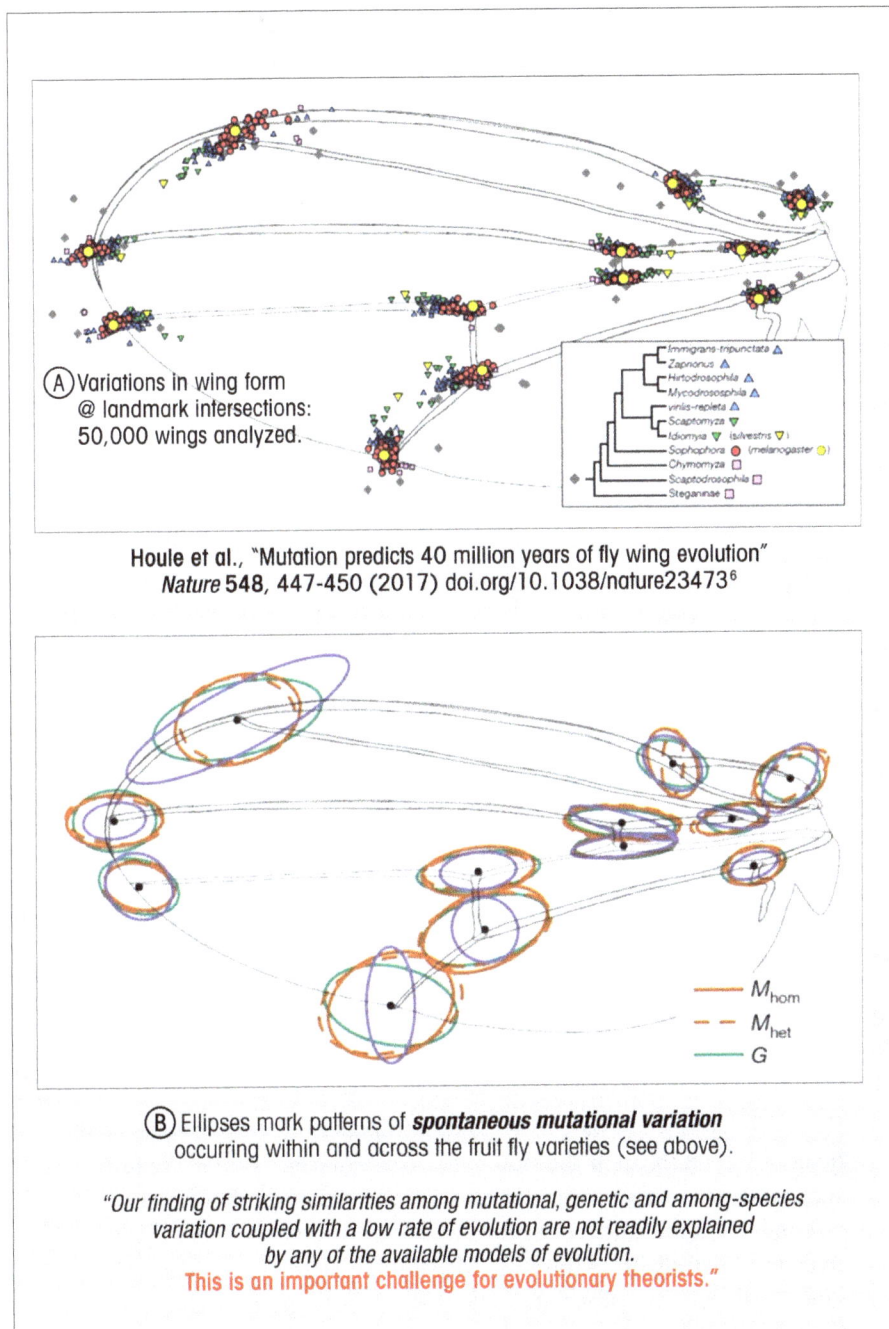

Figure 60. Co·GENESIS & Spontaneous Mutational Variation.

VALIDATION BRIEF: 61

Humans, primates, and several varieties of insects achieve eusociality, the highest level of social order. These species have overlapping generations with cooperative care of the offspring. Typically, there is a division of labor, evident in the caste system of insect societies, where the ability to perform specific tasks is inborn to segments of the colony. *Figure 61* illustrates these divisions, diagrammatically and literally, in a species of termite (*Reticulitermes*).

In a revolutionary paper published in *Population Ecology* in 2020,[7] Kenji Matsuura put forth his theoretical *genomic imprinting* model to explain the developmental pathways and social behavior of this termite species. In contrast to the DNA model of genomic imprinting (see Kinship Theory, pages 38-39), Matsuura's lab and field studies confirmed that only RNA genomic imprinting aligned with the empirical data. A further contradiction to the Kinship Theory in this species is the replacement of the Queen, which occurs by way of Asexual Queen Succession (cloning). Therefore, no paternal/maternal conflict as described in Haig's Kinship Theory is present.

Matsuura's Assertions

"...empirical data and mathematical modeling revealed that genomic imprinting played fundamental roles in the evolution of termite reproduction systems and also implied its association with the origin of eusociality"

"We are now at the dawn of a paradigm shift in the study of social evolution, having the extraordinary opportunity to uncover the epigenetic inheritance and genomic imprinting from both mechanistic and evolutionary points of view."

By 2020, I was well aware that major advances in the life sciences were being achieved through the study of *model organisms*; simple life forms, such as the nematode, fruit fly, and (in this case) termites, to provide fast and economical access to the evolutionary and molecular dynamic that underlies all life. However, Matsuura's placing epigenetic inheritance and genomic imprinting (key RNA biological mechanisms of Co·GENESIS) at the center of his assertions on social evolution, constitute a wholly unexpected and critical validation of this aspect near the end of the writing of this book.

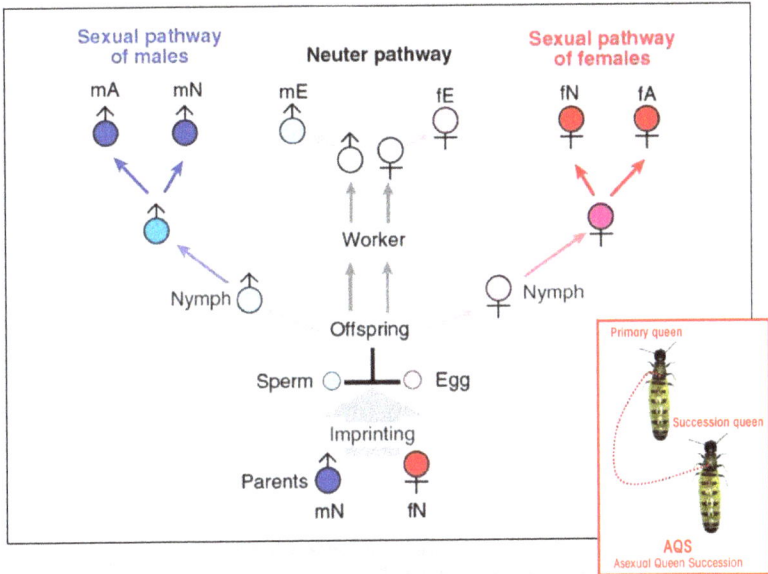

Matsuura, Kenji,
"Genomic imprinting and evolution of insect societies"
Population Ecology 62.1 (2020): 38-52. doi.org/10.1002/1438-390X.12026.[7]

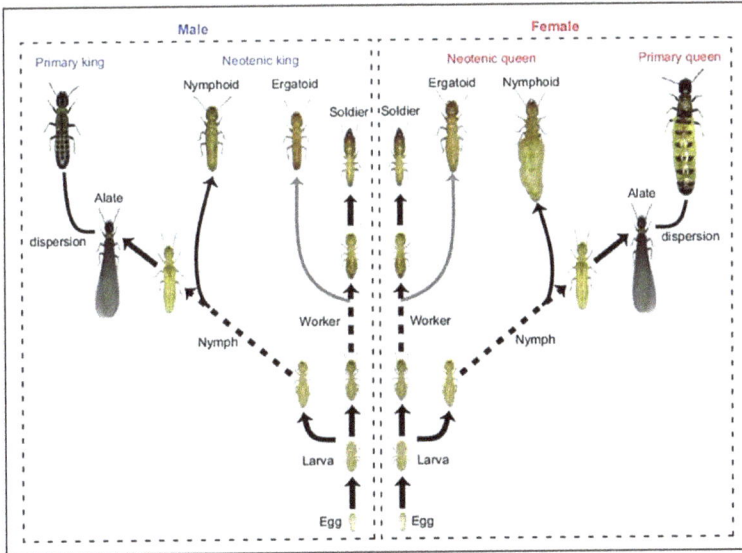

Figure 61. GENOMIC IMPRINTING: Social Evolution in Termites

VALIDATION BRIEF: 62

Never in my wildest dreams did I think that a research team would publish key validations bearing on *Two Minds* and *Co·GENESIS* using such explicit parallels and experimental method. The timing could hardly be better.

The Paris Brain Institute, Linneweber's et al., under the direction of Bassem Hassan published "A neurodevelopmental origin of behavioral individuality in the *Drosophila* visual system" in *SCIENCE* in March of 2020,[8] finding not only the brain/behavioral link to individuality but also demonstrating that the genetic inheritance of the BODY is separate from the epigenetic non-heritable expression of the MIND.

Assertions: Gerit Arne Linneweber et al.

"The origins of behavioral individuality are a central question in neuroscience, psychology, and evolution. Here we establish a link between [left/right] variability in the development of the brain and the emergence of individuality of animal behavior."

"This can serve as a robustness factor for both the individual and the population by increasing the chances of survival of any given genome [species] in case of strong selection pressure [environmental threat]."

"We speculate that similar mechanisms and consequences will hold true in other species, including humans."

Researchers placed a fly on a small island surrounded by water, within a uniformly illuminated translucent cylinder, with only two opposing and unreachable black vertical stripes in the visual field, "inducing the fly to walk back and forth between them…" See *Figure 62*. Male and Female flies demonstrated walking patterns ranging from a highly object-oriented pattern between the stripes, to a randomized non-object-oriented pattern. By "repeated breeding of parental animals with a specific behavioral trait" (e. g., the most object-oriented), they could not change the wide range of epigenetic variation in the perception/behavior of offspring even after seven generations. The DCN neurons in the fly brain were isolated as the source of a left/right wiring asymmetry. In our Theory, this variability flows from the epigenetic mechanism of Co·GENESIS: RNA imprinting/RNA primacy—the RNA MIND.

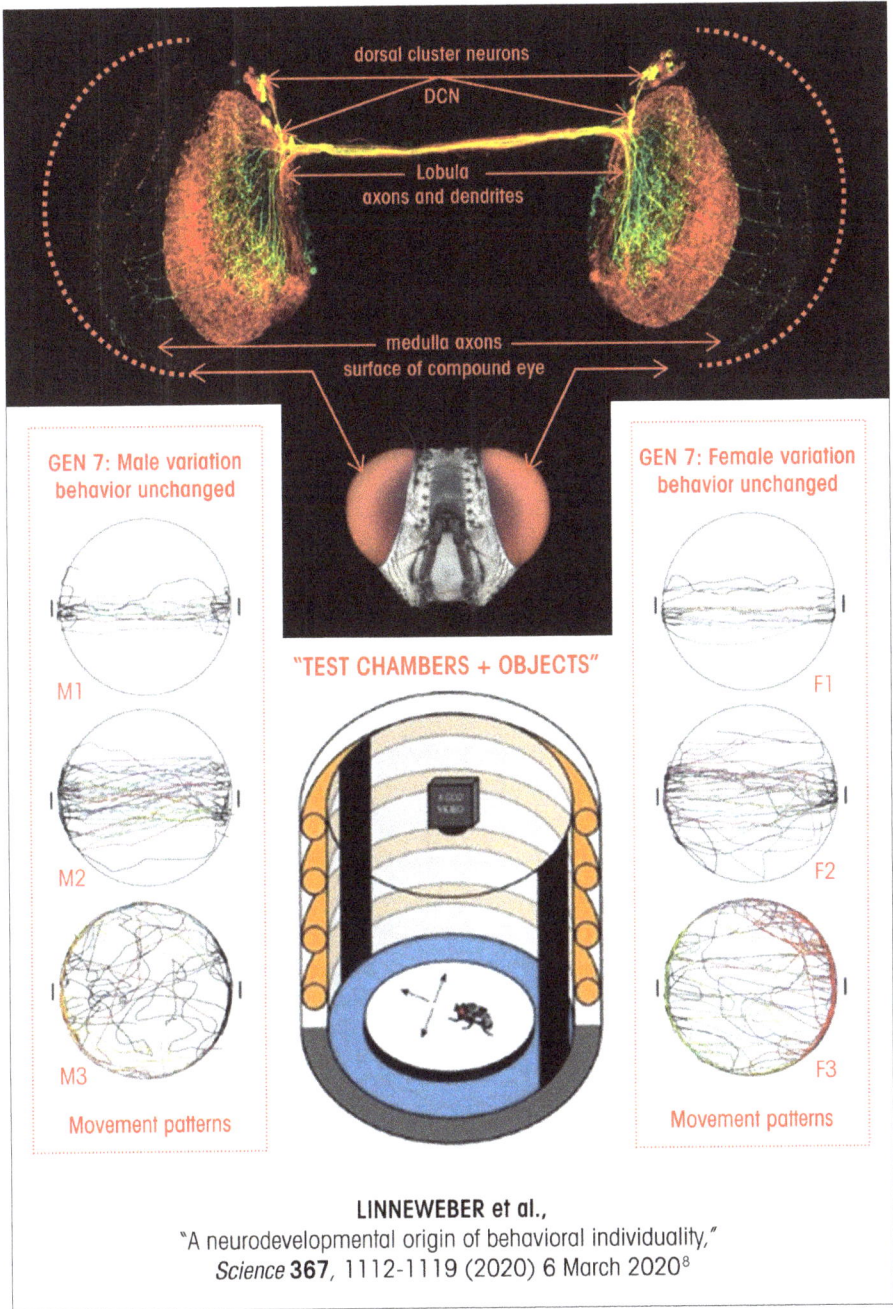

GEN 7: Male variation
behavior unchanged

M1

M2

M3

Movement patterns

dorsal cluster neurons

DCN

Lobula
axons and dendrites

medulla axons
surface of compound eye

GEN 7: Female variation
behavior unchanged

F1

F2

F3

Movement patterns

"TEST CHAMBERS + OBJECTS"

LINNEWEBER et al.,
"A neurodevelopmental origin of behavioral individuality,"
Science 367, 1112-1119 (2020) 6 March 2020[8]

Figure 62. BEHAVIORAL INDIVIDUALITY: Separation of Body and Mind

BRIEF: 63

Although animals come in a wide variety of shapes and sizes, the binary intelligence that informs their Minds, typically results in the binary (bilateral) shapes of their Bodies. See *Figure 63*. Only 1% of biological intelligence goes into making this vast array of physical forms while 24% goes into their regulation and operation. Mendelian genetics and the Modern Synthesis may describe this assemblage of protein, but the larger task of imparting perception, behavior, consciousness, and social order to assure survival requires the remaining 75%: the complex, often viral, and poorly-understood regions of Dark Matter: Co·GENESIS and the RNA Mind.

In Chapter 6 we stepped up in scale to examine the evolution of the RNA Mind across generations, its response to the environment, and the curation of memory. In Chapter 7 we will examine the larger social, and cultural significance of RNA primacy and the basis for a fatal misunderstanding at the scale of our species. Epigenetic and genetic variables from the Male and Female lines of descent are cast and recast by RNA because, metaphorically speaking, Nature's objective is not a "normal" Mind, but the variation of all Minds. A variable that has a positive impact in one casting, may have a negative impact in the next. Misunderstanding this binary dynamic of our evolution is placing our existence at risk. Specific evidence of the danger we face, how we can take constructive action to address it, and the broader implications of Co·GENESIS are the subjects of Chapter 7: "Significance and Meaning."

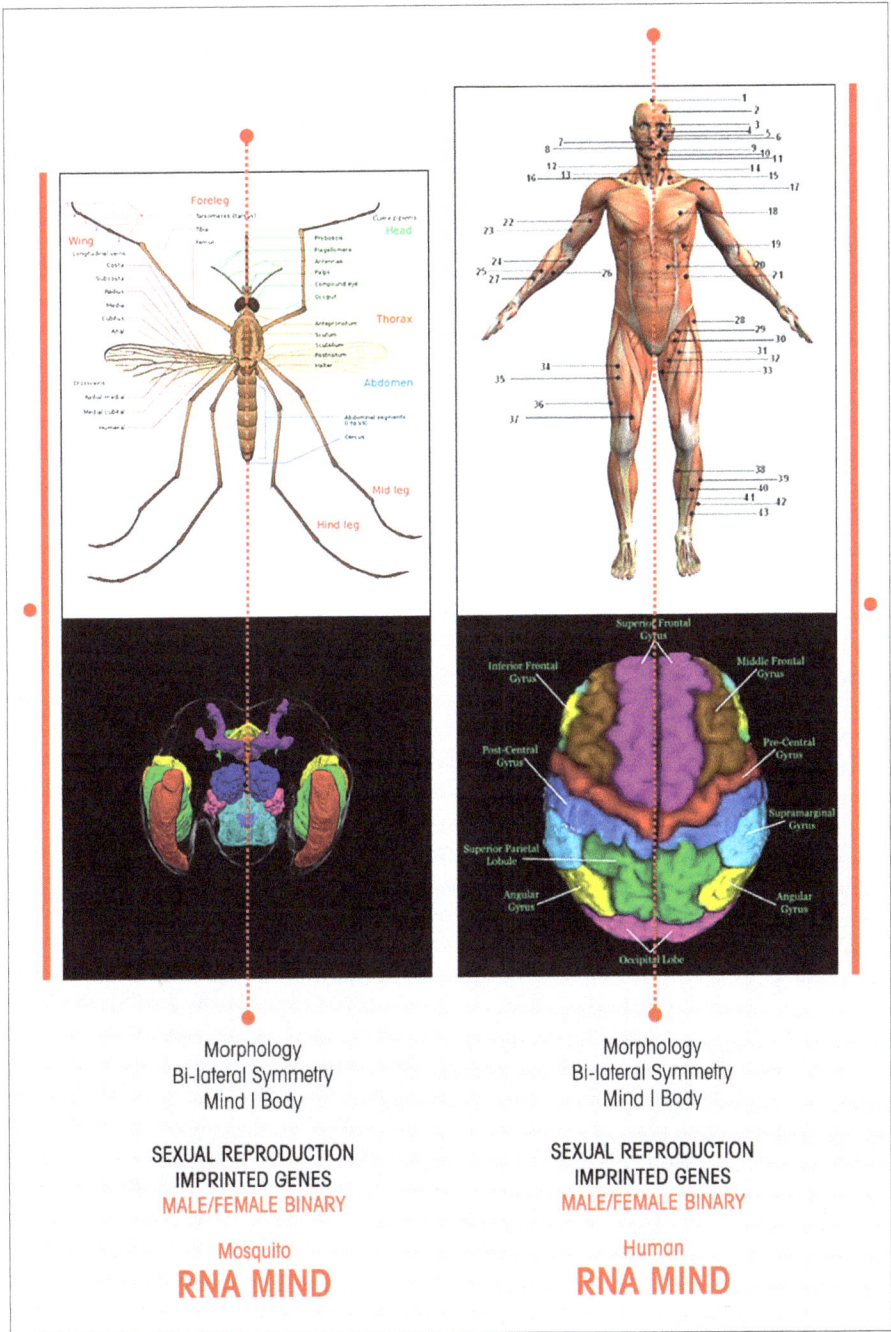

Figure 63. The UNIVERSALITY of Co·GENESIS and RNA Primacy

CHAPTER SEVEN

SIGNIFICANCE & MEANING

A central thesis of Co·GENESIS is that complex organisms, including humans, have evolved through sexual reproduction and the RNA binary code. The first single-celled organisms achieved *homeostasis* (balancing internal conditions to survive changes in the external environment) by incorporating an internal energy source—mitochondria.[1] However, achieving homeostasis in complex multi-cellular life required much more: the processing power of a binary code accomplished through sexual reproduction and derived from male-female dichotomy (differentness) ±1 billion YBP. This early form of binary biological intelligence evolved beyond homeostasis to the variability of perception/behavior across species, and most recently, to deliberative consciousness in humans. Clarification of this transition to complex life forms can be found in the recent work of researchers seeking a bridge from single cell division to sexual reproduction and who, in the process, found support for the inheritance of adaptive characteristics.

Biologists studying the origins of sexual reproduction—prior to the evolution of sexual organisms and gametes—have succeeded in identifying a primordial candidate, complete with its environmental context and biological mechanisms. Aurora Nedelcu, a biologist at the University of New Brunswick in Canada studies multicellular algae, the *Volvox* species, that reproduce asexually by simple cellular division. When placed under environmental stress, whether high heat or the lack of nitrogen, these simple unary cells fuse (effectively having sex) to become binary spores, a more resistant configuration in adverse conditions. The researchers found that when they removed the stressors from the environment, Volvox reverted to unary cells and restarted their asexual reproductive cycle. Not only did this cellular dynamic allow Volvox to survive through hostile conditions, but the species emerged from such events in an improved, more hearty and resilient form.

The April 23, 2020, issue of *Quanta Magazine*[2] reported on Nedelcu's research and the linkage of Volvox to sexual reproduction via cellular merging (similar to gametes joining in the zygote: *embryogenesis*), and the reverse process of separation (similar to binary cells transitioning to unary gametes: *gametogenesis*). The Volvox cellular division in asexual reproduction is known as *mitosis*; the reverse process from binary-to-unary, is *meiosis*. The latter is the most significant for our reference, the moment of the transfer of adaptive information.

Volvox foreshadows the transition from asexual to sexual reproduction (See *Figure 64*). The "parental" cells came together to survive harsh environments as spores, but incidentally they exchanged information and passed on adaptive resilience to subsequent generations—a primordial argument against the Weissman Barrier and in support of Darwin/Lamarck. In short, the Volvox asexual analog points to an opening, an evolutionary pathway to the inheritance of adaptive characteristics, sexual reproduction, and the Code of Life. One billion YBP is our working assumption for this start of sexual reproduction and the RNA MIND. Sponges, fish, amphibians, plants, insects, reptiles, mammals—life as we know it—followed.

Moving beyond the DNA model should not be wholly unexpected, given the outcome of the Human Genome Project in 2001–2003.[3] Many scientists expected 100,000 to 120,000 DNA (protein-coding) genes to reflect the complexity of life, but found only 20-25,000. Many organisms, including moss and microscopic water fleas, have more. And yet scientists still focus on the limited 1% of protein-coding genes that appear to follow Mendelian Genetics. Consistent with Co·GENESIS, many men (high maleness) and many women (high maleness) in the sciences are advantageously predisposed to be: <u>logical, analytical, and reductionist—the essence of the scientific method, which has delivered stunning advances for the world's population.</u> However, the primarily non-protein-coding domains of Dark Matter—the RNA MIND, must address whole system complexity and future uncertainty. Embedded at the following levels in the Order of Life are complex products of sexual differentiation, the biological mechanism that makes muti-cellular life possible.

1. Co·GENESIS explains sexual reproduction and the resulting binary code that makes the regulatory control of multicellular life *(homeostasis)* possible. A seemingly simple example is the on-going *variation* (±) to maintain a critical human set point: 98.6° F, thermal balance.

2. The RNA MIND is the biological intelligence (Code of Life) of a species at a moment in time. Through adaptive evolution of the Body as it seeks equilibrium with the environment, the MIND passes down as a series of species, redefining the concept of species boundaries, *(plasticity-ambiguity)*.

3. The Applied RNA MIND is the unique neurological/endocrine expression of perception/behavior of the individual within a species. Maximum variation achieves resilience of the species (social order) in the face of future unknowns. The RNA MIND, metaphorically speaking, is employing random variation honed by natural selection to address future events: *uncertainty.*

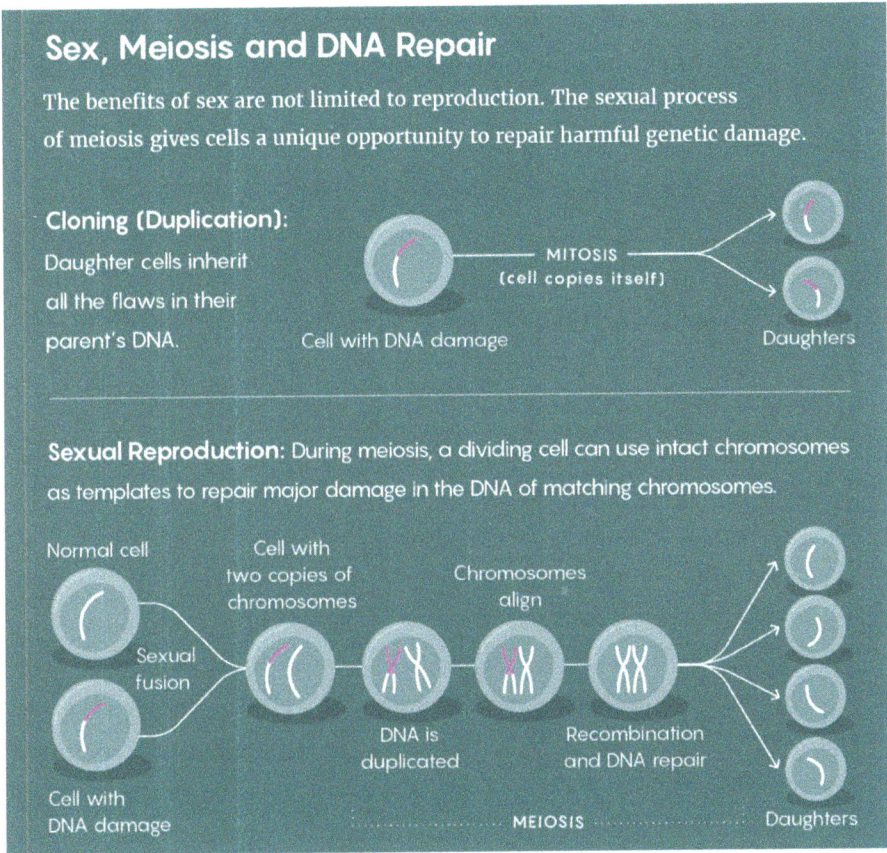

Figure 64. ASEXUAL vs. SEXUAL: Reproduction & Adaptation

The Theory of Co·GENESIS in the academic sense is not just a shift from a DNA-centered to an RNA-centered biology, but rather a reconceptualization of the processes of evolution—most closely aligned with Charles Darwin's Pangenesis hypothesis. In this new context, Humankind stands more firmly on common ground because the male and female life experiences of our parental lines inform the Mind of every Human. While variation in the way we each see the world, our social order, creates endless debate and even conflict, it is the delicate balance that assures our viability as a species, Nonetheless, Co·GENESIS will be upsetting existing norms and revealing risks which have been ordered here under three headings: Social/Cultural/Political, Scientific/Academic, and Humankind at Risk.

Social/Cultural/Political

There are currently two dominant views on the nature of the sexes. Both are wrong. The first is the conservative/traditional view of men as masculine—the assertive, physically strong, confident providers for their families, and women as *feminine*—the deferential, nurturing, and empathetic centers of family life and community called "normal" by believers—other variations are characterized as abnormal, or worse, immoral. The other extreme is the modern/liberal view that males and females are absolutely equal and that sexual and gender stereotypes projected by parents and the media from early childhood prevent the recognition of equality. The idea that a given behavior may be more characteristic of men or women, is an affront to their view of the world. Both of these extreme societal viewpoints fail to explain the maleness/femaleness spectrum we observe in perception-behavior across the population, much less its essential role in the social order, the viability of Humankind.

Co·GENESIS explains the variation of Human perception-behavior that we witness every day: the spectrum of instinctive-subconscious personality and sexual attraction (see *Figures 27 & 28, pages 49, 51*) interacting with the overlay of deliberative consciousness. Through consciousness, Humans moderate the pliable aspects of their behavior for social integration with other mindsets they encounter/select in life. The universal characteristics identified here are *maleness* (e.g. analytically intelligent, assertive, decisive, etc.) and *femaleness* (e.g. emotionally intelligent, creative, empathetic, etc.). No male or female can be judged on their blend of these two qualities in a first meeting, even though men have a greater number of high maleness ratios and women have a greater number of high femaleness ratios. These opposite poles of the binary Mind are present in each of us, blended in near-infinite combinations at the joining of the hemispheres.

The uniqueness of each brain is easily demonstrated by the *amygdala*, a key binary organ of perception/behavior located to the left and right of the hypothalamus and connected by 40 million of the 250 million nerve fibers linking the hemispheres. In the upper half of *Figure 65*, the internal amygdala is shown in front and side views of the brain. The *corpus callosum* and related cross-connections, which have 210 million nerve fibers is also described in the lower half of the illustration, visible here within the transparent surrounding hemispheres.

As we encountered with speech and language (see *Figure 58, page 103*), the left hemisphere amygdala is the analytical/action perception of the threat/aggression memories from our past, while the right hemisphere amygdala carries the avoidance/defensive perception attached to those events. While the term deliberative consciousness suggests that there is time for deliberation, the amygdala is the rapid-reaction center for action in the face of a threat to survival.

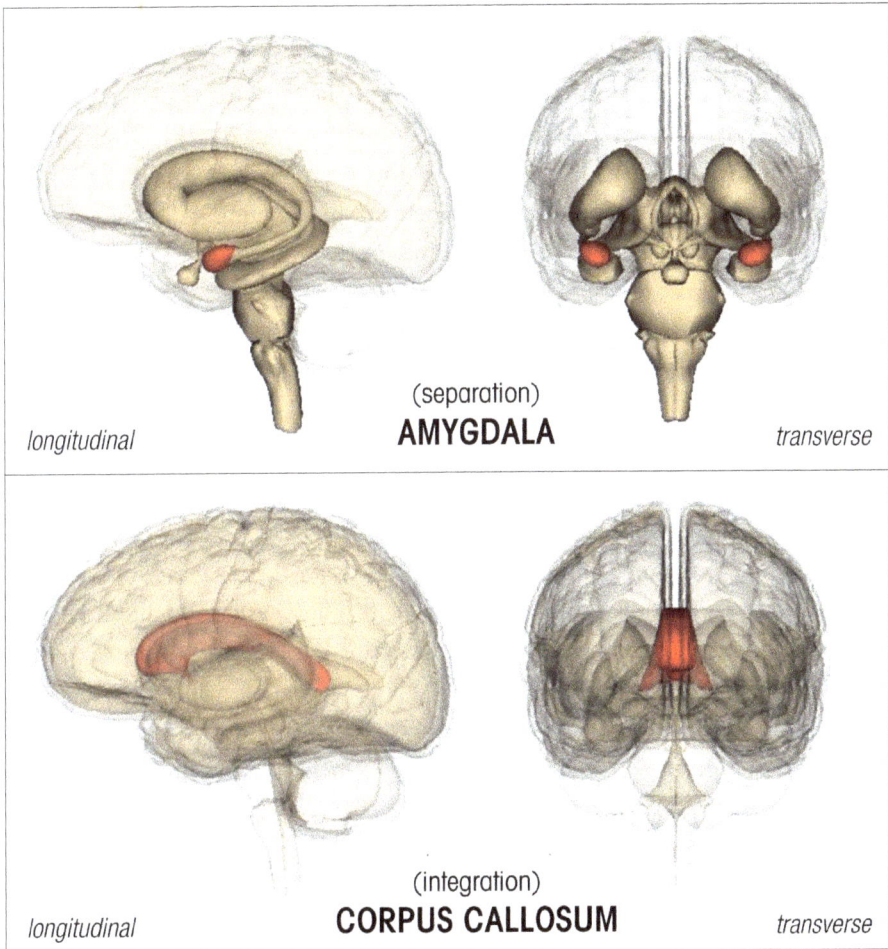

(separation)
AMYGDALA
longitudinal   *transverse*

(integration)
CORPUS CALLOSUM
longitudinal   *transverse*

Figure 65. SEPARATE MEMORY SYSTEMS: in Deliberation

The power of this duo acting at the subconscious level is such that, in the face of an extreme threat, before conscious awareness, they can "hi-jack" the brain into instant action, (e.g. side-stepping an oncoming car). Consistent with our assertions, many researchers have referred to the left and right amygdala as having "separate memory systems" working together to resolve a threat. Actions range from aggression (maleness FIGHT) to retreat (femaleness FLIGHT), the end points in the rheostat of our reactions. A high maleness female facing an attacker with a knife may choose to fight; a high femaleness male may choose to flee. Either decision, or both, may turn out to be correct or incorrect, it just depends. Different reactions to life-threatening situations create a resilient population. Anatomical gender is most certainly not the determinant of perception-behavior.

An allegory for the variability imparted by Nature is "the Fort." Surrounded and outnumbered by the enemy who will be attacking at dawn, we expect every soldier to be brave and fight to the death to defend the Fort. Let's call that *Scenario 1*. However, in *Scenario 2*, most soldiers do just that, but one has collaborated with the enemy, a couple of others sneak out overnight between enemy lines, and some, seeing their forces being overwhelmed, hide among the bodies of their fallen comrades, or surrender in a final effort to survive. Nature, metaphorically speaking, sees *Scenario 2* as, by far, the more resilient species. A single "normal" view of the world means that it is only a matter of time before a unanimous fatal decision is made—the opposite of a resilient population.

Variations for left bias of the amygdala, the Grand Central Terminal of threat assessment/reaction, assures that an optimum segment of men and women will always be assertive soldiers, police, and firefighters who will run to danger in defense of the common good, while a right bias assures men and women, who tend not to go into harm's way, and possess worldviews of empathy and non-confrontation. Both segments, as we will see, constitute the competitive and strategic mix of perception-behavior necessary for the long-term survival of the species.

Given that every brain appears superficially identical, how is it possible that such a wide range of expressions arise in every generation? On July 23, 2021, a visually striking answer was published in *Frontiers in Neuroanatomy*.[4] Researchers in Seoul, South Korea developed a rare whole-brain analysis by tracking the three nerve fiber types connecting the hemispheres, assessing their density ratios (terminations per 1 cubic mm), and assigning density-dependent primary colors to map the cerebral cortex of individuals. Using the dataset of the NIH Human Connectome Project[5] and the Brainnetome Atlas,[6] which details the interior subunits of the brain, the cerebral cortex images were developed for twelve individuals in *Figure 66*. The top exterior brain is shown on the left; the internal cortex (mapping of subunits) is on the right of each pairing. This individual differentiation is achieved with an economy of means. The basic genetic brain is passed down to all humans, but the wiring is epigenetic, binary, and unique. Individuals B, F, G and I stand out...*what could be going on in those minds?*

The unique instinctive-subconscious Minds (personalities) are cast, and the stable expression of social order is maintained. Evolution to an optimized balance of inter-individual difference is achieved at the smallest unit of epigenetic/genetic expression, the individual nucleotide "SNPs," which can snap into a number of alternative, often predisposed, alignments.

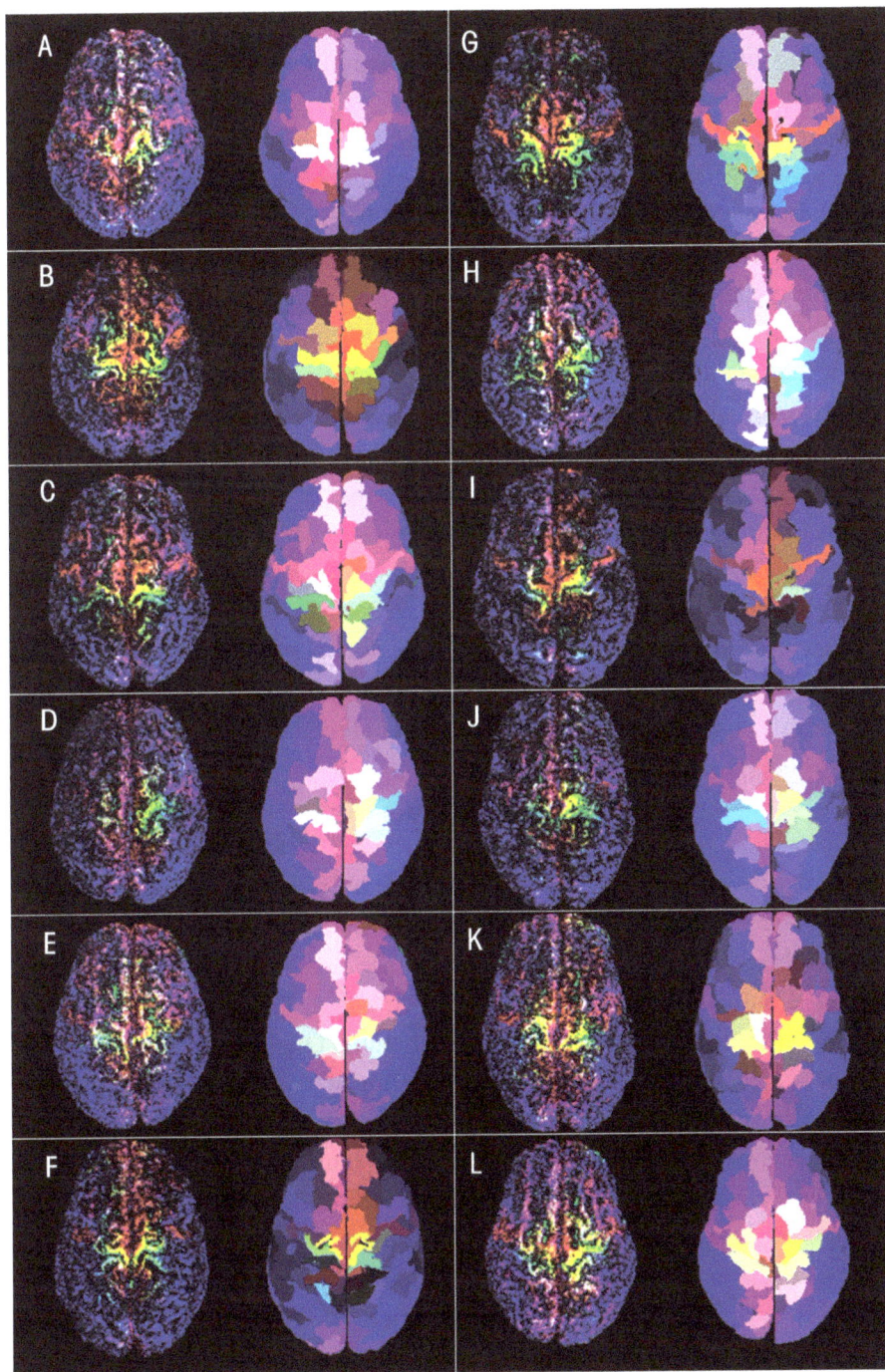

Figure 66. BRAIN MAPPING: Track-Density Ratio of Twelve Individuals

While the randomness of individual outcomes is likely the result of a code instruction (an algorithm) for variation, the biased pathways of the SNPs and other variables, such as lines of Code that can "jump" to alternate locations, (*transposons* and retrotransposons) are all Dark Matter players in maintaining the larger pattern of the species. Similar to 5-card draw Poker, each hand dealt is random, but there will statistically be the same number of pairs, flushes, straights, and full houses overtime. Once again, we see a biological process needing a binary code for the ideal mix of heroes, cowards, independents, conservatives, liberals, etc.

The Theory of Co·GENESIS asserts that Humankind is far more inter-woven than the DNA model; we stand on deeply shared biological ground. All Humans start with a universal Body that six weeks into embryonic development is neuter, neither male nor female, and we all possess an Applied RNA Mind which, no matter your anatomical gender, is a male and female ancestral legacy in collaboration. *Figure 67* is a researchers' abstraction of the forest of nerve fibers which, with a minimal expenditure of metabolic energy, is uniquely wired for each individual's left/right Human consciousness.[7] The corpus callosum, shown in bright red-orange, also serves as the bridge for speech, deliberation, and our inner voice. Anatomical gender, sexual orientation, and race are not visible here because this image is the biological universal of Humankind. In this measure, we are all equal.

The philosophers of the Enlightenment, as interpreted by Thomas Paine and later, Thomas Jefferson in the Preamble to the Declaration of Independence, recognized the equality of Humans and the Universal Human Rights of "Life, Liberty, and the Pursuit of Happiness." These were deemed inalienable, that is to say, inherent to the person. John Locke, the British philosopher, identified them as "Natural Rights." Central to the American and French Revolutions, such rights inform the concept of self-governance. The model of American De-mocracy draws on the United States' spectrum of mindsets for a diverse and representative government, one in which open debate and deliberation lead to decisions on the conduct of the nation. In short, Democracy can be a re-creation of the complete Human Mind and deliberative consciousness which, through the on-going election of diverse Mindsets, can become the Mind of the Nation. The greater the accuracy of representation, reflecting the actual diversity of minds within the Nation, the more effective the Democracy. By supporting individuality of initiative and personal expression, balanced by a commitment to the common defense and common good (social security, healthcare, education, etc.), a nation worth fighting and dying for is born—a foundational aspect of national security.

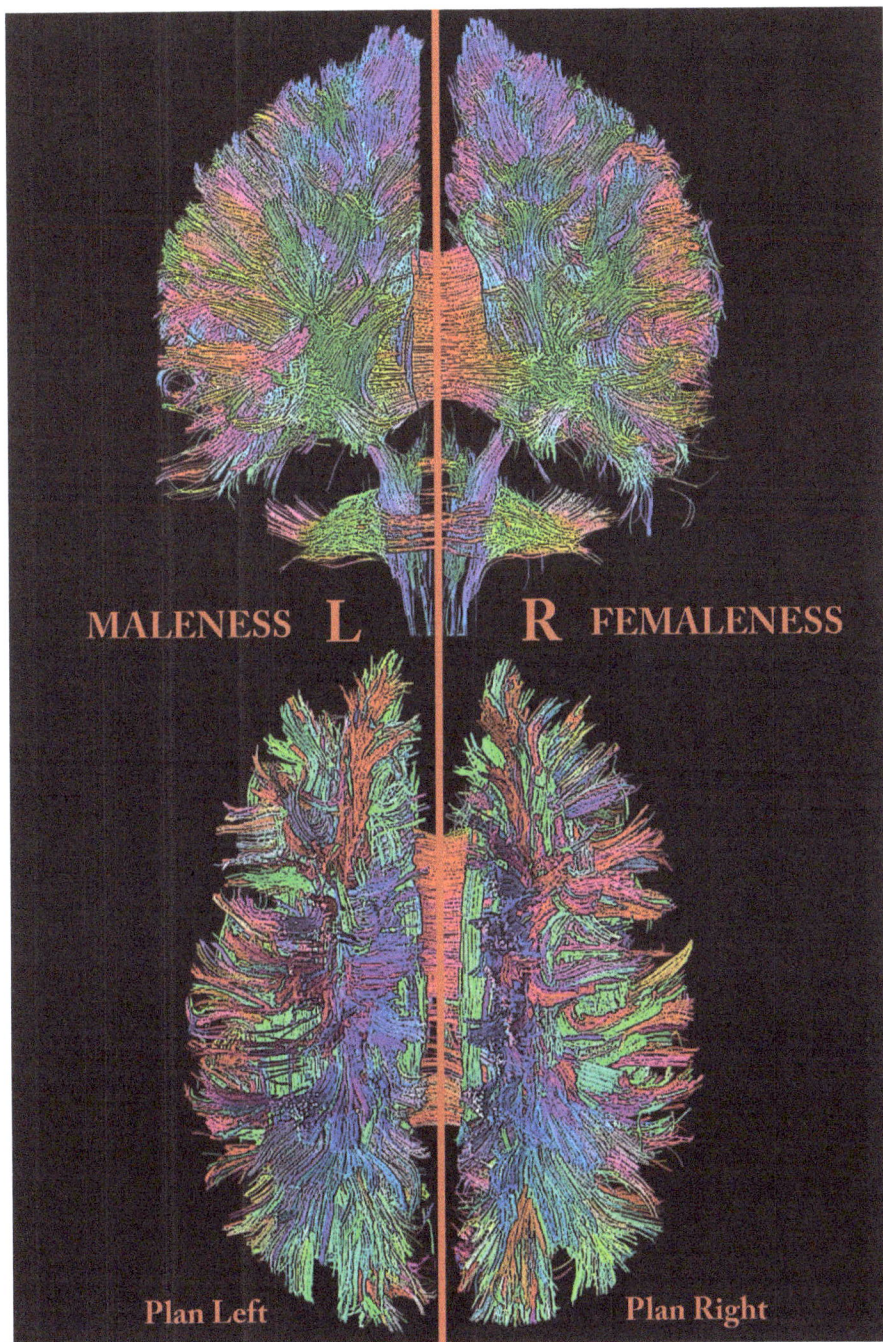

Figure 67. The UNIVERSAL RNA MIND: HUMANKIND

By contrast, authoritarian governments, whether led by dictators or like-minded committees, fail to mobilize their greatest national asset: the diverse Minds of their people. No level of threats, bribery, weaponry, or training can create the commitment to fight and die for the country's leaders if they are seen as corrupt and self-dealing (e.g., Iraqis vs. ISIS and Afghans vs. Taliban). The international flow of human capital (immigration) and financial capital (safe-havens and investment) confirm the draw of individual freedom and human rights; capital consistently flows from authoritarian extremes to stable democracies.

While the proportions of perception-behavior (e.g., conservative-threat vs. liberal-opportunity) in a population remain stable, their distribution within a nation's political parties can change dramatically. Sitting in my grandmother's living room in Lincolnton, North Carolina in July of 1952 at age 8, I was a minority voice among grandparents, parents, aunts, uncles and cousins as we marveled at the first televised national coverage of the Democratic and Republican Conventions. The Democratic Convention came second and when Adlai Stevenson concluded his acceptance speech, I announced (to the abject horror of all present) that if I could vote, I would vote for Eisenhower. Abrupt corrections were served up, followed by the well-worn phrase: "Honey, you are just not old enough to understand." Of course, this was a veiled reference to Abraham Lincoln and the Civil War...Republicans! In 50's America, there was the calming effect of the rural demographic being split between the two parties. However, Richard Nixon's 1968 "Southern Strategy," Newt Gingrich's "Contract with America" in 1994, the "Tea Party" in opposition to Obama in 2009 and Trump's 2016 "MAGA" amplifications, have aligned the rural south with the rest of rural America: a near-perfect rural/urban contest. A 100% rural contingent, dominantly maleness/left hemisphere (males and females) is far more aggressive than its dominantly female-ness/right hemisphere urban counterpart, a formula for maximum conflict.

Scientific/Academic

My first encounter with the biological conception of the Human Mind was the brilliant book, *Consilience*, by Edward O. Wilson (*Figure 68*).[8] Although neither I, nor he (I believe), recognized it at the time, Wilson was seeking a unity of knowledge between the two sides of the Mind: the Sciences (rational, analytical, *maleness*) and the Humanities (emotional, creative, *femaleness*). In *Figure 68* (in red) I have added my characterization of those fields of knowledge as an extension of the Human Mind. Only in Minds approximating a 50/50 balance of these two oppositional expressions (see *Figure 28*, page 51) can such a bridge be built—Edward O. Wilson, being an exemplar of the first order.

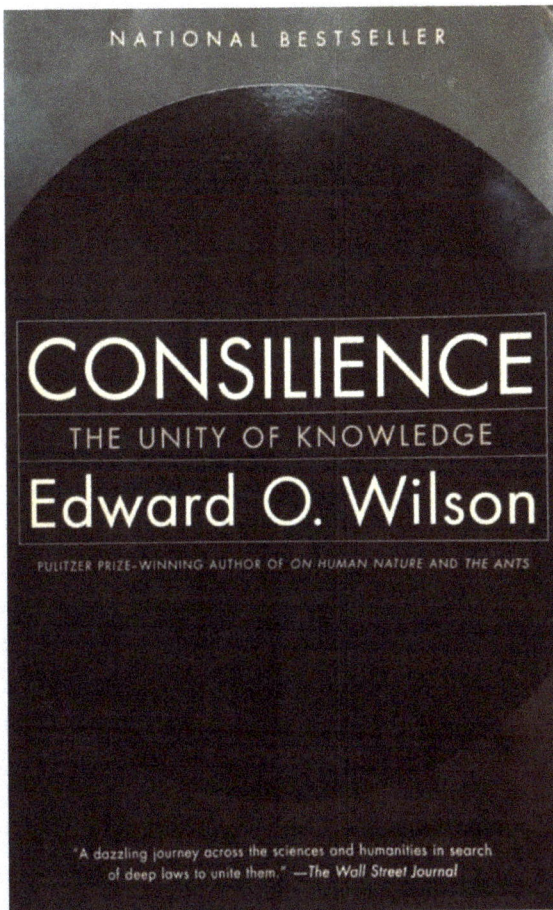

NATIONAL BESTSELLER

CONSILIENCE
THE UNITY OF KNOWLEDGE
Edward O. Wilson

PULITZER PRIZE-WINNING AUTHOR OF *ON HUMAN NATURE* AND *THE ANTS*

"A dazzling journey across the sciences and humanities in search of deep laws to unite them." —*The Wall Street Journal*

1998

"The greatest enterprise of the mind has always been and always will be the attempted linkage of the Sciences and Humanities."

analytical and reductionist | holistic and integrative
unemotional and mechanistic | humanness and emotion
rote and methodological | innovative and creative

Deliberative Consciousness

Figure 68. FOUNDATIONAL SCHOLARSHIP

In retrospect, my immediate interest in *Consilience* traces back to 1954 and a childhood story I shared in the Foreword of *Two Minds* (2015). My fourth-grade teacher, Mrs. Marjorie Brown, observed that I was a rare combination of a student good in Math and Science (*maleness*) and Art (*femaleness*), and asked if I had considered being an architect. *Consilience* re-ignited my interest in the underpinnings of that question, and after many years of personal exploration, led to the Theory of *Co·GENESIS*.

Four years after the double-helix 3-D model of DNA in 1953 by Jim Watson and Francis Crick—which was aided by the x-ray analysis of DNA structure by Maurice Wilkins and Rosalind Franklin—Francis Crick delivered an iconic lecture at University College London. The lecture proposed a relationship between the information-carrying organic molecules DNA and RNA, and the building blocks of the Body: Proteins. This lecture is said to have altered the logic of biology. I have adopted the diagrammatic simplicity used by Crick in his lectures to describe the flow of information (see *Figure 70A* as published in *PLOS Biology*[9]) to compare the RNA-centered flow of information under Co·GENESIS.

The enlargement of *Figure 70B* shows Crick's proposed one-way information flow, DNA-to-PROTEIN (since found not to occur) and his remaining classic DNA-to-RNA-to-PROTEIN. Crick asserted his Central Dogma of biology as follows: "Once information has got into a PROTEIN it can't get out again." The dotted line represents a reverse flow from RNA to DNA (see *Figure 15*, page 29). Circular-shaped arrows indicate DNA and RNA copying themselves.

In *Figure 70C*, I have illustrated Co·GENESIS in which RNA has interactive information flows with DNA and Protein in forming the RNA MIND, which ultimately feeds back information for the next generation through the gametes. However, the actual nature of Co·GENESIS has to be seen as a two-part Male and Female RNA MIND (immortal-evolving), which forms the Individual (temporal). See *Figure 69* below.

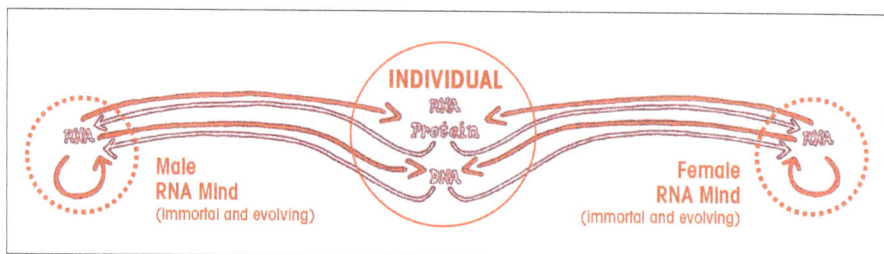

Figure 69. The RNA MIND (immortal) and The Individual (temporal)

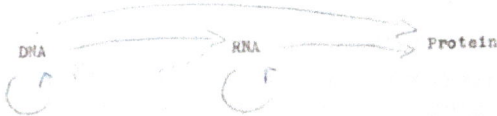

The Central Dogma: "Once information has got into a protein it can't get out again". Information here means the sequence of the amino acid residues, or other sequences related to it. That is, we may be able to have

but never

where the arrows show the transfer of information.

Fig 1. Crick's first outline of the central dogma, from an unpublished note made in 1956. Credit: Wellcome Library, London

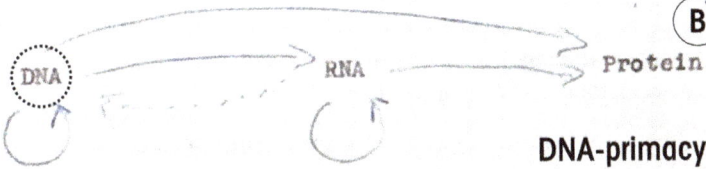

DNA-primacy

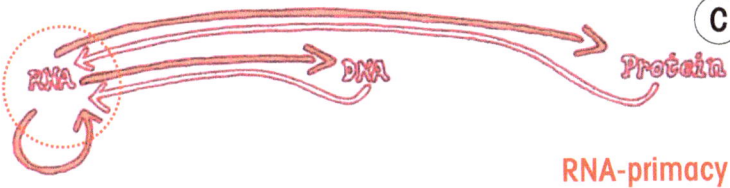

RNA-primacy

Molecular Information Flows: DNA vs. RNA

Figure 70. INFORMATION FLOWS: RNA, DNA, and PROTEIN

The one-way information flow of Crick's diagram served as an affirmation of August Weismann's assertion that adaptive life experiences of the Body (made of protein) cannot pass to the next generation. To the contrary, the Co·GENESIS assertion is: "Once information has gotten into Protein, it can be modified and passed to the next generation." Adaptive characteristics, as described herein, pass from generation-to-generation via extracellular messengers to the gametes, as Darwin proposed, and, as we have further developed, they pass down within an overarching RNA binary order: the evolutionary Theory of Co·GENESIS (see *Figure 71*).

The "random mutations" we observe are likely the most visible parts of an RNA-generated array of outcomes—the optimum range of evolutionarily advantageous individuals possible within the epigenetic and genetic variables retained in the Human genome over the last billion years. Because this is NOT a linear process aimed at a single ideal outcome, but a randomized binary process aimed at variation and the Net Highest Best Outcome, there are going to be a minority of negative results, which tend to be gradually eliminated in the process of Natural Selection. Casting and recasting of variables in each birth results in increasingly ADVANTAGEOUS expressions, and, in some individuals, extraordinary outcomes: math genius, physics genius, artistic genius, musical genius, literary genius, etc. Nonetheless, the same variable that is advantageous in the casting of one individual may be DISADVANTAGEOUS in the casting of others. Variations such as autistic, savant, or bipolar expression would have been selected out long ago, however, they possess key variables in the casting of genius, a massive net advantage for the individual and the species as a whole. For this reason, these elements get reproduced and retained in the genome. Other genetic elements with essential roles we have yet to discover or understand are likely to be associated with many of the remaining, repetitive, genetic diseases. Misunderstanding this inherent characteristic, thinking that these are simple DNA errors that can be permanently fixed, has placed Humankind at the edge of disaster.

Humankind at Risk

In 2017, marking the sixtieth anniversary of Crick's revolutionary presentation, evolutionary biologist, Professor Matthew Cobb from Manchester University published "The lecture that changed biology" in which he observed: "Despite the excitement about what is called epigenetics, which explains how genes can be turned on and off by the environment, this never leads to a change in our actual DNA sequence. Crick's dogma was absolutely right."[10] Cobb's 2017 opinion, it is fair to say, is the majority position of scientists today.

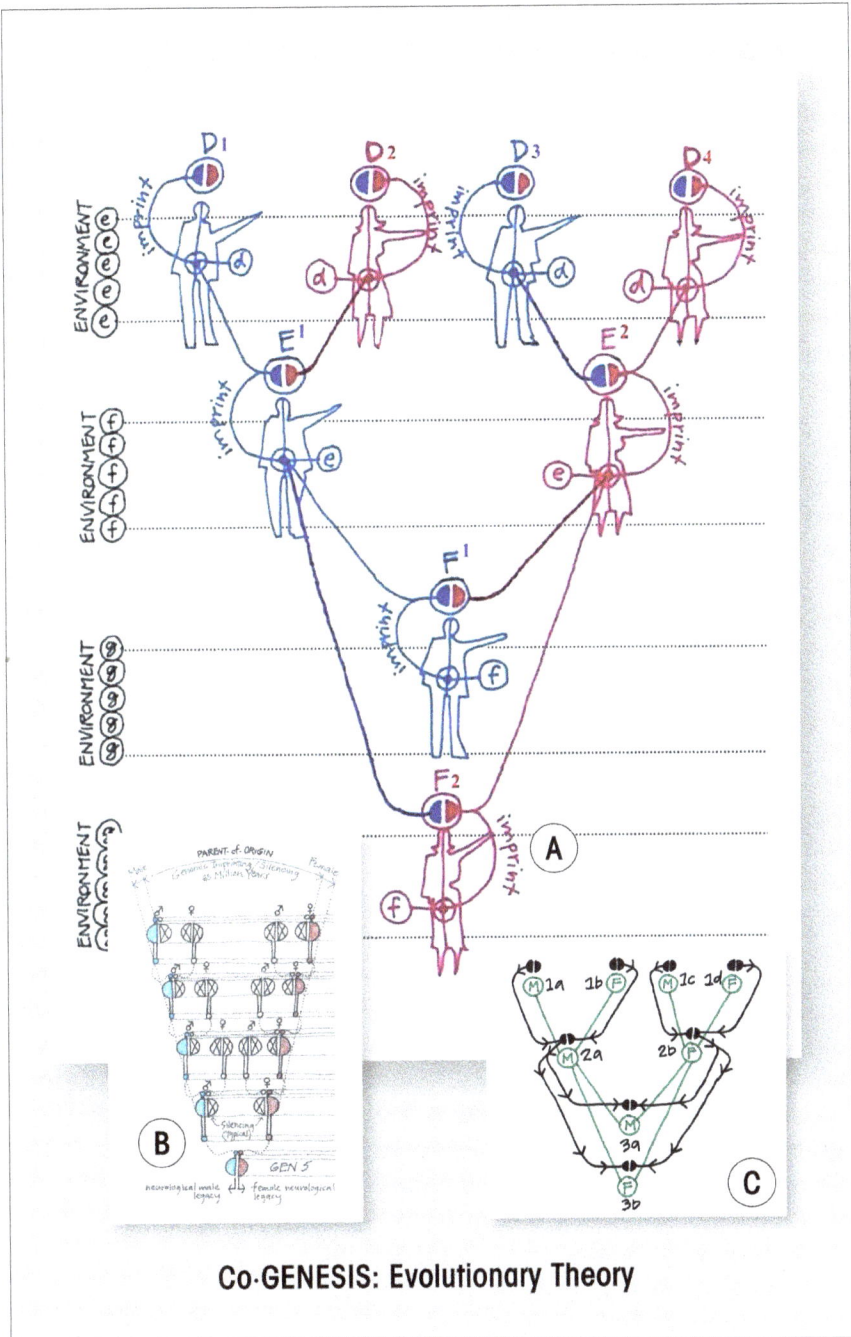

Figure 71. Environmental, Neurological and Mind/Body Relationships

However, an unsanctioned foray into Human DNA germline editing of embryos of non-identical twin girls in China in 2018 has exposed the dimensions of risk and complexity hidden within the DNA model. He Jiankui, a researcher (See *Figure 72*), acted in secret, utilized forged documents, made misrepresentations, and failed to gain fully informed consent. He's intervention has since been determined to be illegal by Chinese authorities and He Jiankui was sentenced to three years in prison.[11]

He Jiankui's stated goal was to protect the girls from their father's HIV by providing them with a variant of the CCR5 gene, which occurs naturally in 1% of Northern Europeans, and imparts immunity to the HIV virus in that population. Nana and Lulu are the pseudonyms for the girls who should have received two copies of the preferred variant with 32 pairs of letters (nucleotides) missing from the gene, instead:

1) Nana had an extra base pair of letters added to one copy of the preferred variant and four deleted from the other copy (*off-target effects*).

2) Lulu had 15 base pairs deleted from one copy and the other copy was unedited (*off-target effects*).

3) All of their body cells were not altered because of *mosaicism*, which occurred when subsequent divisions of the fertilized first cell did not fully pick up the edits, resulting in edited and unedited cells.

The collective effect is that immunity from HIV is highly unlikely, and there are now multiple mutated versions of CCR5, products of human error, that can be inherited by any offspring. Zaria Gorvett, a senior journalist for *BBC.com/future*, noted the CCR5 errors listed above and observed that, in 2018, CCR5 was mostly known for its role in relation to the HIV virus. However: "Today, there's an emerging consensus that it [CCR5] has a variety of functions–including in brain development, recovery from strokes, Alzheimer's disease, the spread of certain cancers, and the outcome of infection with other pathogens."[12] In short, we already have corrupted versions of genes which have a role in brain development—the little-known Dark Matter domain of deliberative consciousness—poised to flow into our gene pool, into the RNA MIND.

We must not modify one more variable in the germline until we understand the tens, or hundreds, or more roles that each variable plays. Scientists now have a vast knowledge of the epigenetic-genetic signatures of disease, and have shown the ability to address the statistically small amount of collateral damage inherent to Co·GENESIS through safer individual extract-edit-return techniques using an individual's soma (body) cells. In this manner, with extreme care, diseases can be addressed, albeit more slowly, without risking the future of Humankind.

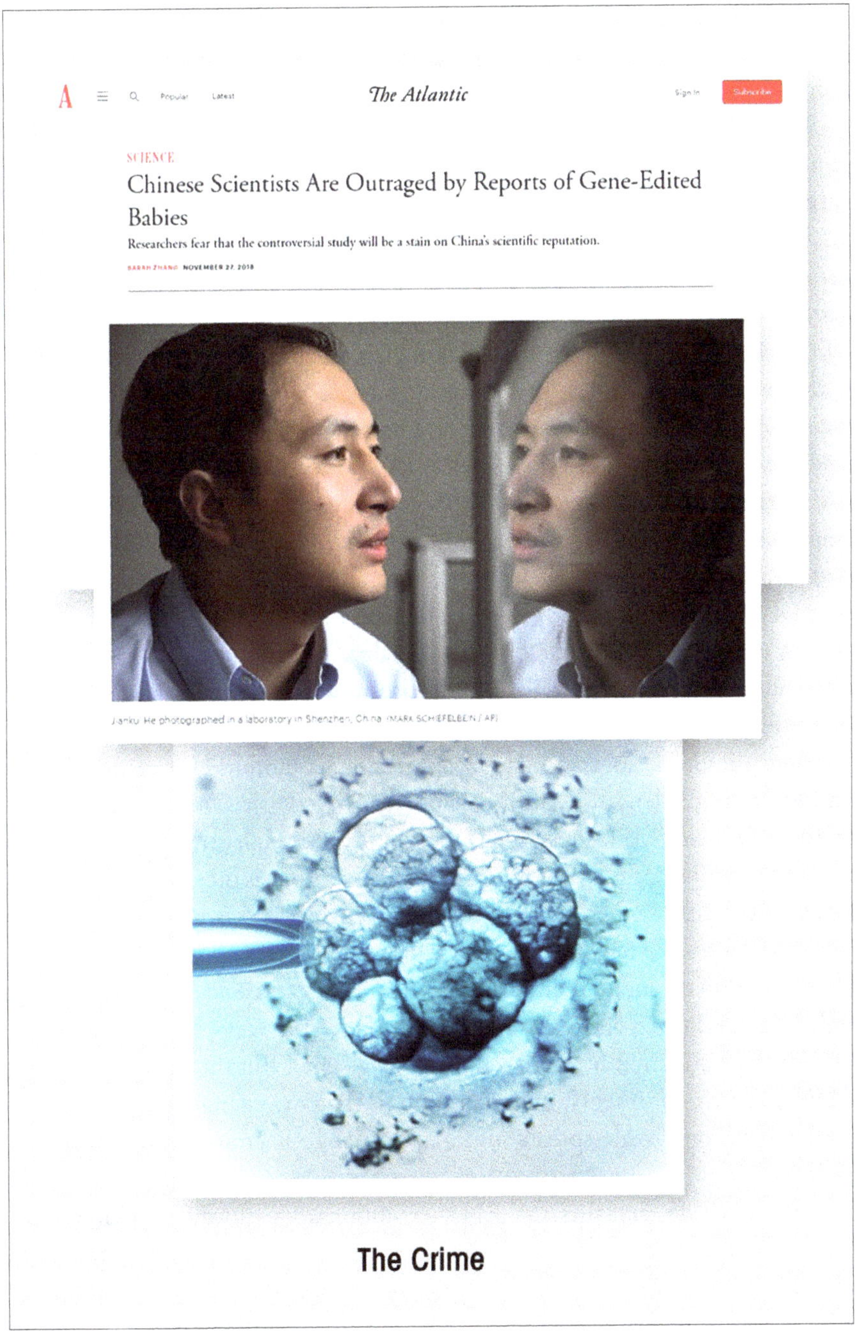

Figure 72. ILLEGAL HUMAN GENE EDITING: He Jiankui

I initially planned to conclude this book with the He Jiankui incident, the condemnation of the perpetrator and the specific risk he imparted to humanity. Never did I anticipate that a brilliant Nobel Prize-winning scientist and a New York Times best-selling author would come together in *The Code Breaker* (Simon and Schuster, 2021)[13] to catapult the issue to the national forefront, wrapped in an articulate and disarmingly benign narrative. See *Figure 73*.

The genius of Jennifer Doudna and Walter Isaacson will assuredly not be diminished by my observations here, which are limited to their public statements on the nature of germline editing and advancing the widely-held DNA model. In *A Crack in Creation* (Houghton Mifflin Harcourt, 2017),[14] Doudna characterizes the course of evolution into modern times as "littered with organisms that certainly didn't benefit from the mutational chaos that underpins evolution." Elsewhere she states: "[Nature's] carelessness can seem like outright cruelty for those people unlucky enough to inherit genetic mutations that turned out to be suboptimal." That Nature is doing a horrible job, and that biomedical science can do it better, is a widely-held belief, rooted in the simplicity and omissions of the DNA Model. Another pro-germline editing argument is that the extreme suffering of those with genetic disorders, such as Huntington's Disease, demand immediate action. Isaacson cites the disease multiple times in underlining the urgency of the current situation.

Henry T. Greely, an ethics and law professor at Stanford University, addresses germline editing at length in his recent book *Crispr People: The Science and Ethics of Editing Humans* MIT Press (2021).[15] Greely, although strongly condemning the criminal actions of He Jiankui, sees germline editing as inevitable, but needing adequate safeguards which he sees as being available. In his generally reassuring public statements, Greely has observed that we don't seem to be as worried about our unintentional genetic modifications—like the increase in our genes to digest starch, which followed civilization's Agricultural Revolution—as we are about our intentional genetic modifications (a false equivalency). But the most dangerous of his DNA-centric comments, in my opinion, is that any change in genetic editing would not become wide-spread "unless the change is enormously, enormously beneficial."

Co·GENESIS explains the RNA evolutionary pathway to Human consciousness—a masterwork—in continuous response to the environment over a period of a billion years. While Huntington's Disease affects 1 in 37,000 births, a germline genetic or epigenetic human error can affect 37,000 out of 37,000 births in a family line. As we will see, even when deliberative consciousness is impaired, reproduction can frighteningly continue on. In order to stepback from the precipice and avoid the risk to future generations we need to more fully understand the decisions that have led us to this moment and take action.

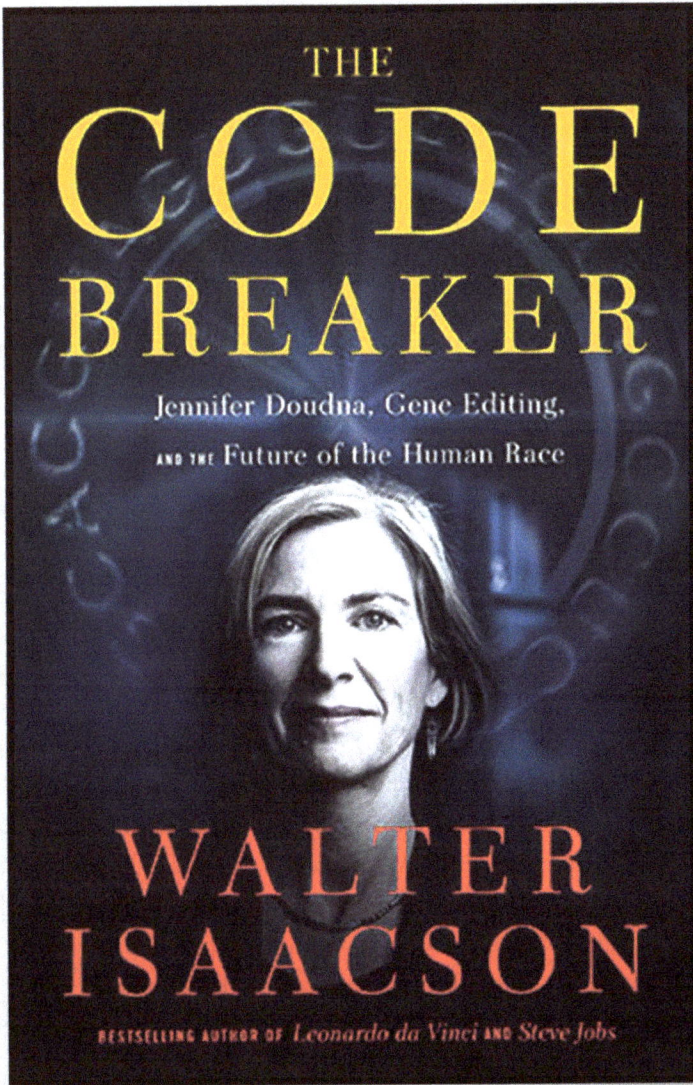

Figure 73. GENE EDITING and the Public's Perception

Human consciousness, having evolved through language and its integrative role across the hemispheres, separates us from members of the genus *Homo*, and all other life forms, which react through instinctive-subconsciousness. See *Figure 74*. In Nature, there is a phenomenon called *co-evolution*, which we see in the competitive evolution of cheetahs and gazelles. The process of natural selection rewards the fastest cheetahs, and the most wily/deceptive gazelles, as they co-evolve each other, predator and prey. However, the handoff of these increasingly adaptive advantages must wait for the next birth. By contrast, deliberative consciousness allows us to evolve information in our Minds. Through access to the male and female legacies (our inner voice) we perceive, deliberate, and evolve in real time. We read the writings of others and advance them, we hear music and evolve it to new forms, we see art and architecture, and counter with our own. Humans are constantly engaged in the competitive challenge and co-evolution of each other. The world into which we are born is embedded with the co-evolutionary competitions of our ancestors, through narratives, books, art, architecture, music, science, customs, values, beliefs, etc.

How do we know that consciousness is the dividing line of life on Earth? All we need to do is examine the 100,000 years before and after the divide. Human consciousness emerged 100,000 YBP ago in Africa and the out-migration of Humans began about 60,000 YBP. What followed was the first-time extinction of the members of the genus *Homo* and the Human domination of all habitable continents by 12,000 YBP. The Agricultural Revolution followed, plus the Babylonian and Egyptian civilizations, Plato's Theory of Mind and Body, Darwinian Evolution, the Enlightenment, the Atomic bomb, Moon landings, and 7.8 Billion Humans (36% of the mammalian biomass) controlling domesticated livestock (60% of the mammalian biomass) with a remnant of wild mammals (4% of the mammalian biomass)—the Anthropocene Age.[16] What happened in the 100,000 years prior to the divide? The thinly dispersed genus *Homo* advanced their stone-chipping technique for tools and weapons.

I return to the subject of deliberative-consciousness because it is, in my opinion, our most vulnerable attribute which is being placed at inexcusable risk when we modify the biological pathways of its evolution. Without it, we are not Human, it is everything to our species. The uncomfortable reality is that our knowledge about the risks and consequences of germline editing is dependent on the use of experimental animals (mammals) in our labs, mammals that do not have deliberative-consciousness, mammals that cannot speak.

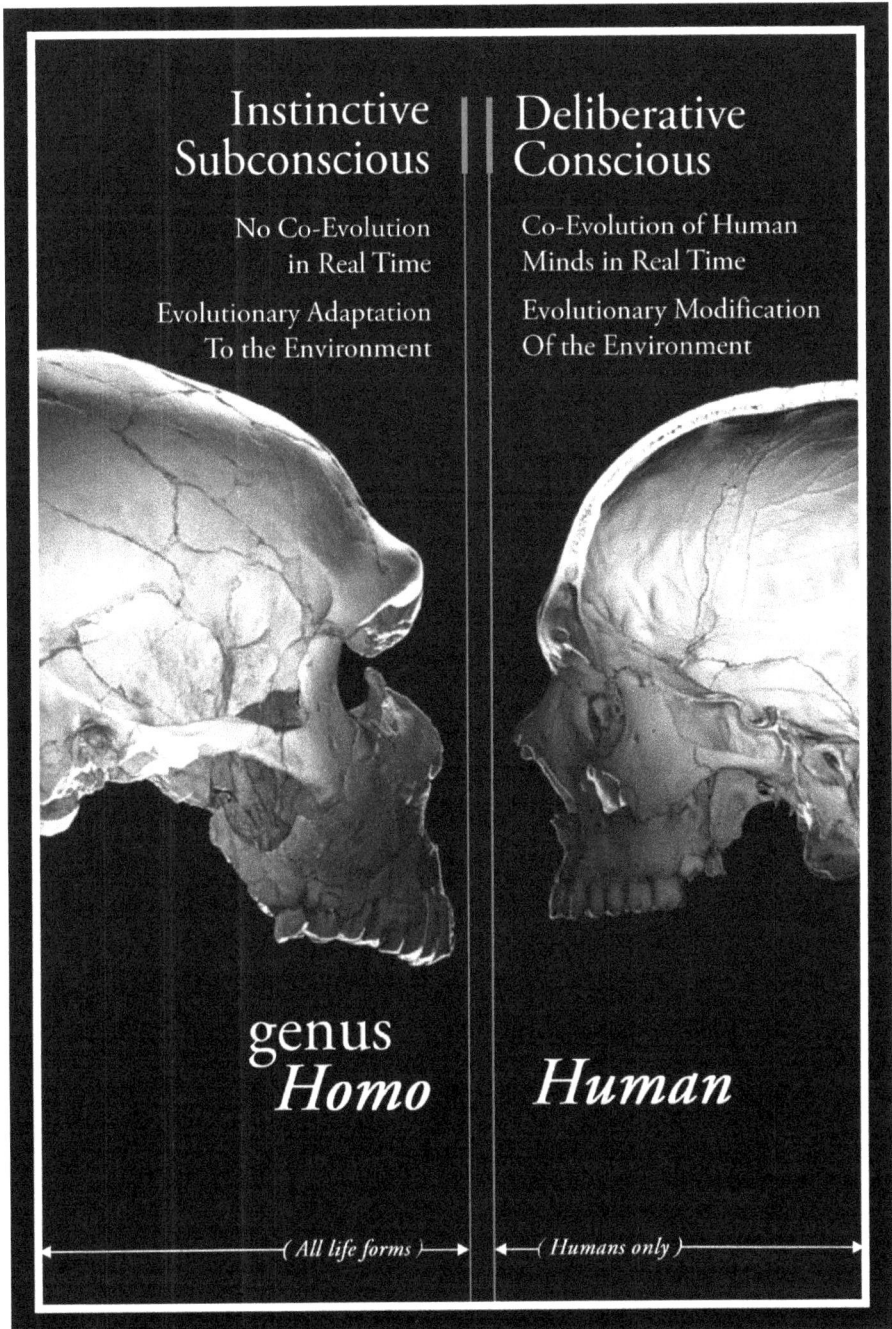

Instinctive Subconscious || Deliberative Conscious

No Co-Evolution in Real Time

Evolutionary Adaptation To the Environment

Co-Evolution of Human Minds in Real Time

Evolutionary Modification Of the Environment

genus *Homo*

Human

(All life forms) → ← *(Humans only)*

Figure 74. INSTINCTIVE SUBCONSCIOUS vs. DELIBERATIVE CONSCIOUS

The chronological array of mammalian brains in *Figure 75a* includes many of the animals used in laboratory experiments (mice, rats, rabbits), the livestock we breed for our consumption (pigs, cows), and domesticated pets (cats, dogs) bred for our companionship.[17] Although the chimpanzee is our closest living relative, for economy, ease of handling, and short life span, the mouse wins out as the Human stand-in for genomic experimentation and risk avoidance.

Figure 75b is the Mammalian Genome, the Code of Life, written in stable DNA molecular form with all the information for RNA, DNA, Proteins, in other words, the RNA Body and the RNA MIND.[18] The 1% of the genome (protein-coding), shown here in purple, is the center of the universe for the DNA model and long thought to be the only functional part, while the remainder was "junk." However, following the completion of the Human Genome Project in 2001-2003, and the finding of significant non-coding RNA functional elements, the ENCODE project was initiated to identify the remaining functional elements in the mouse and Human Genome. The blue and gray circles, shown expanding out to occupy over 80% of *Figure 75b*, represent the portion declared to be "functional" by ENCODE in 2012, based on biochemical activity. The blue/gray also represents the maximum extent of the "Dark Matter" (still disputed as to whether it is functional or junk). However, it is the red circle labeled "*Ultra-conserved*," primarily within the Dark Matter, that opens a window into the risks we are now taking.

As we observed with miRNA, the conservation of genomic sequences over long evolutionary timeframes typically indicates high functional importance. Within all mammals there is this red circle of extreme evolutionary sequence conservation, primarily consisting of RNA sequences in the Dark Matter domain, <u>that is identical across human, mouse, and rat genomes</u>, dating back to our last common ancestor 65-80 million years ago. This grouping is referred to as "ultraconserved" and was the subject of the paper "Deletion of Ultracon-served Elements Yields Viable Mice" published in *PLoS* (2007), by the team of Ahituv et al.[19] The surprising finding was that the deletion of four of the ultraconserved elements apparently had no effect. The researchers stated that their studies: "ruled out the hypothesis that these constraints reflect crucial functions required for viability." Because it came in the wake of the Human Genome Project and the discovery of the unexpected functionality of RNA, the study provided support for the "junk" proponents, characterizing Dark Matter and RNA as inconsequential.

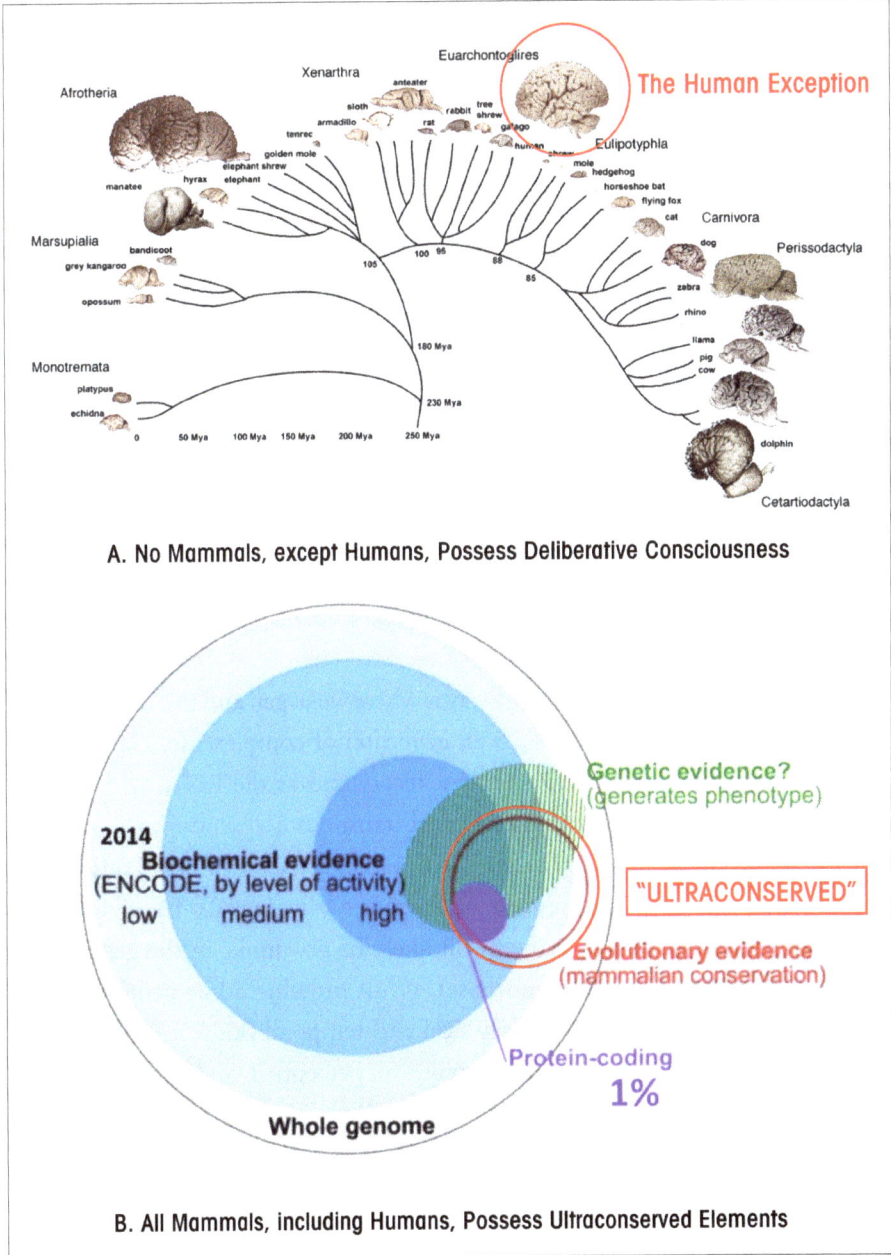

A. No Mammals, except Humans, Possess Deliberative Consciousness

B. All Mammals, including Humans, Possess Ultraconserved Elements

Figure 75. EXCEPTION and COMMONALITY within Mammals

Snetkova et al., publishing in *Cell* (2018) and *Nature Genetics* (2021), returned to the matter of the ultra-conserved elements (UCE) See *Figure 76a*.[20] Digging deeper into the subjects of the aforementioned paper by Ahituv et al., Snetkova's team heroically reversed the finding of "...no crucial functions." I had intended to use their new conclusion in support of Co·GENESIS, but quickly realized that they were pointing to something much larger. Drawing on the work of Heckler and Hillier, published in *Gigascience* (2020),[21] in aligning the ultra-conserved elements of sixteen species of vertebrates—illustrated in *Figure 76a*—we see that the species: fish, amphibians, reptiles, birds, and mammals, are in the order of the rise of vertebrates. The date of the last common ancestor of zebra fish and Humans (±400 million YBP) and the date of vertebrates moving from ocean to land (±380 YBP) support this as an orderly progression of life forms. The ultra-conserved elements are forming the common "trunk" supporting the expansion of information in the subsequent and increasingly complex species. Consistent with the theory of Co·GENESIS, the trunk is the pathway of the RNA MIND: lncRNA, miRNA, and the RNA imprinted genes of the Dark Matter region.

An RNA transgenerational highway is where we began and the UCE provide a compelling signature for just such an evolution of complex life. Dark Matter may well be neither junk nor dark when recognized as the RNA MIND—the curated workshop where transposons, retrotransposons, pseudogenes, introns, exons, etc., (retained as essential by the on-going process of assimilation) are being cast and recast in the process of forming individual Codes of Life. For this reason, we can expect that there will likely be no "junk" in the genomes of Humans (every fragment with a purpose). While protein-coding genes form the most visible product of RNA, the physical and temporal Body, it is just "meat" without the RNA MIND adapting the physical, perceptual, and behavioral characteristics to the environment—the essence of life in all its forms.

To prove their point, Snetkova et al. (2021) examined 23 ultraconserved enhancer elements. However, the key for our purposes is element *hs122*, which interacts with the *Arx* gene associated with the developmental phases of the brain. By attaching a blue-indicator, the developmental actions of Arx in the forebrain are visible (see *Figure 76b*). The researchers substituted an *hs122* with 5% of the nucleotides mutated (See *Figure 76c*), seeking to determine if the enhancer is functional. The resulting outcomes could range from no change, to loss, or to gain of function (see *Figure 76d*).

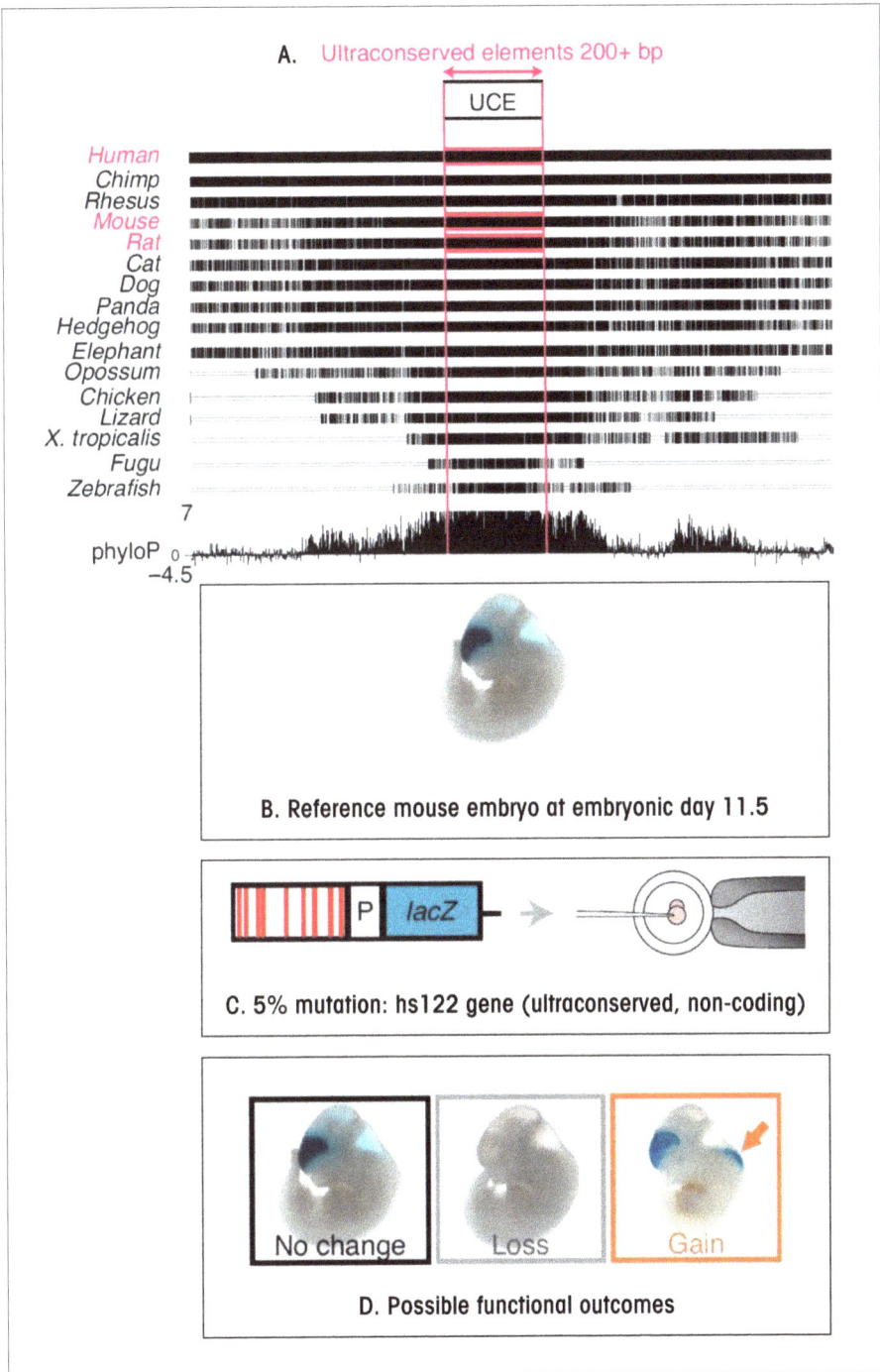

A. Ultraconserved elements 200+ bp

UCE

Human
Chimp
Rhesus
Mouse
Rat
Cat
Dog
Panda
Hedgehog
Elephant
Opossum
Chicken
Lizard
X. tropicalis
Fugu
Zebrafish

7
phyloP 0
−4.5

B. Reference mouse embryo at embryonic day 11.5

P   lacZ

C. 5% mutation: hs122 gene (ultraconserved, non-coding)

No change   Loss   Gain

D. Possible functional outcomes

Figure 76. VALIDATION of ULTRACONSERVED ELEMENTS

The outcomes reversed the hypothesis that the ultraconserved elements are not necessary by demonstrating that a 5% mutation of nucleotides in one of these elements, *hs122*, had the terrifying outcome described in *Figure 77*. Embryonic brain development was stunted and deformation of the *hippocampus* and reduction of the *dentate gyrus* occurred: all three are foundations of deliberative consciousness in Humans. See *Figures 77c & 77d*. Three key observations follow:

1. The ultraconserved elements of the RNA MIND are essential for life.

2. The experimental elimination of ultraconserved enhancers in 2007 appeared to have no effect because researchers used body weight, longevity, ability to reproduce and litter size to determine viability. The mice had neurological defects which were later described as "*subtle effects*," not easily picked up in the lab. *The mice lived on to reproduce and pass on their defects of Mind—Human Minds and the Social Order are at risk.*

3. The human RNA organic molecules and genomic elements have evolved to an ideal societal balance of variation. The binary Code of Life delivers deliberative consciousness to Humans with a minimal number of non-functional outcomes which can be addressed without modifying the germline. No editing of the germline can be allowed given our lack of knowledge and the existential threat to Humankind.

Snetkova et al. observed that many of the other 23 ultraconserved enhancers they mutated did not show obvious impacts. But importantly, they noted that there may be "additional yet-to-be-identified, regulatory or non-regulatory functions of these sequences contributing to their extreme conservation in evolution." In the four years since He Jiankui's 2018 announcement of his germline gene editing, Crispr cas-9 (Cc-9), has been reexamined:

1. The US National Institutes of Health's Somatic Cell Genome Editing Consortium (SCGE) publishing in *Nature* (April 2021) takes a firm line: "germline editing is not only excluded as a goal but is considered to be an unacceptable outcome that should be carefully prevented." [22]

2. Publishing in *PNAS* (April 2021), the team of Alanos-Lobato et al. at the Crick Institute report: "Unintended genome editing outcomes [Cc-9] were present in approximately 16% of the Human embryo cells analyzed...Our work underscores the importance of further basic research to assesses the safety of genomic editing technique in human embryos." [23]

A *Consensus Statement* by 29 scientists in the June 2023 issue of *nature reviews molecular cell biology*[24] challenges the role of the 1% 20,000 protein-coding genes, the basis of DNA Primacy. The >100,000 long non-coding RNA (lncRNA) are called out as genes—found to be expressed by cell type, tissue type, and developmental stage of life and to drive cellular regulation, individual variation, and fitness. In short, in my reading, the Statement affirms RNA Primacy.

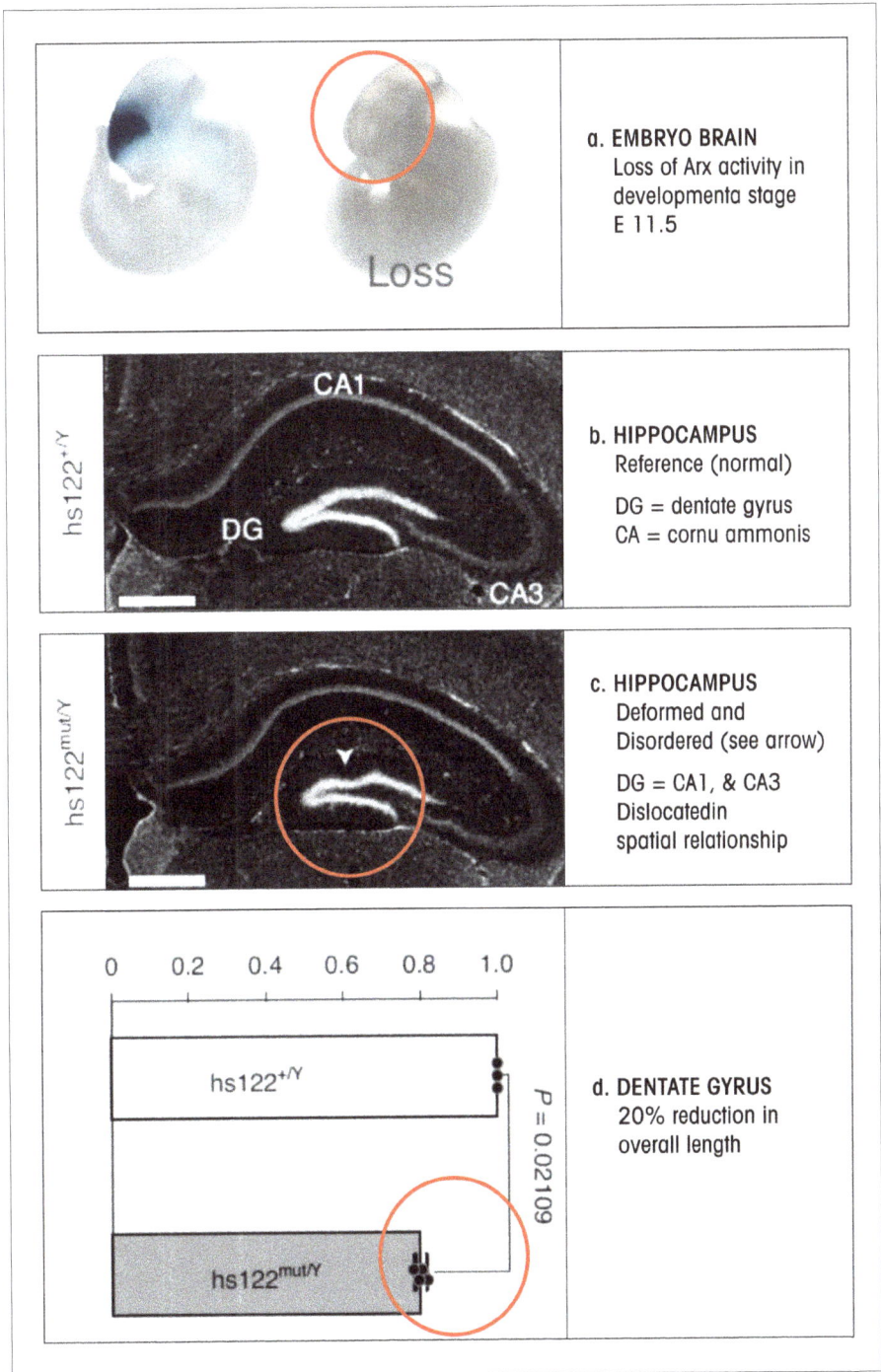

a. EMBRYO BRAIN
Loss of Arx activity in developmenta stage
E 11.5

b. HIPPOCAMPUS
Reference (normal)

DG = dentate gyrus
CA = cornu ammonis

c. HIPPOCAMPUS
Deformed and
Disordered (see arrow)

DG = CA1, & CA3
Dislocatedin
spatial relationship

d. DENTATE GYRUS
20% reduction in
overall length

Figure 77. EVIDENCE of BRAIN and MIND IMPACTS

As I write these words, the biomedical community, via patents and billion-dollar investments, are introducing edited and active-editing genomic material into Humans that may hit the reproductive line—this in addition to the international threat of new molecular knives promising fame and fortune to scientists and investors by directly modifying the germline. We are about to destroy our greatest gifts—consciousness and the social order. Humans must not directly or indirectly modify the Code of Life. Urgent actions are needed:

I) An international moratorium on human heritable germline modifications, with severe criminal penalties;

II) An international moratorium on insertion of cellular genomic editors with germline risk to humans, with severe civil penalties;

III) Convening of an international council of cross-disciplinary experts to openly debate and propose a consensus pathway forward that manages risk to the future of humankind—voted on and adopted by all nations.

With a commitment to act, we can confront this most immediate, and essentially unseen, threat to humankind. In closing, it is appropriate that I address how and why I, an architect, came to write this book. As I mentioned, my pursuit of human-centered design placed me within the disciplines of the life sciences, primarily the neurosciences, psychiatry, and evolutionary theory. Immediately, to my eyes, there were interconnecting relationships that were not being explored. Certain areas of research were not funded, or the significance of narrowly-defined projects were being missed in the larger context. The inherent concerns of grant-making and academic institutions (public image—cultural and political) were marginalizing important areas of study. I believed that as an independent observer, from an integrative discipline, I could make advancements in this critical field of study. Over the past twelve years, I have read many thousands of research papers documenting the evolving knowledge base of the life sciences. I am left with the highest regard for the intellect, rigor, and honesty of these teams and the, near-universal, selfless observations on the limits of their own work. Without funding or lab, I was free to follow points of interconnection, able to shift focus and disciplines in the continuous pursuit of an understanding. Co·GENESIS, I believe, is the advancement I sought. I trust that my interpretations and assertions, (looking over the shoulders of many others) have been reasonable.

The concluding image, *Figure 78*, is a reminder of the individuality of our view of the world afforded by Co·GENESIS. The biological interaction of Music and the Mind, blends with the specific experience of a time and place, positioning them within the fabric of our unique memory.[25] No one has ever seen the world as you see it and, through offspring and/or co-evolutionary influence, each of us bends the arc of the RNA MIND of Humankind.

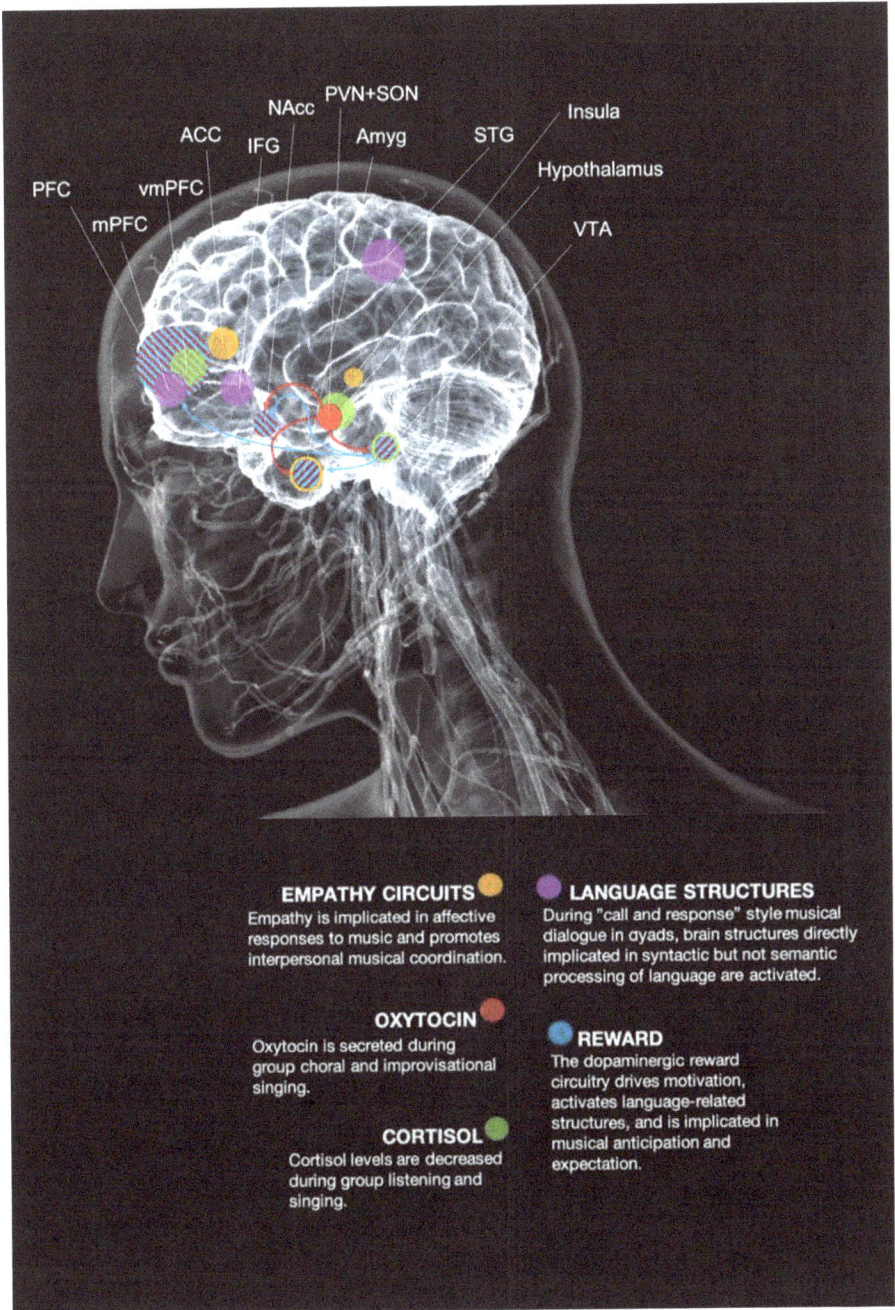

Figure 78. BIOLOGICAL INTERACTION of MUSIC and MIND

AFTERWORD

Consciousness reversed the priority of Nature in Human evolution. For a billion years sexually-reproducing life was "wild": instinctive-subconscious mind-sets. Life was experienced as simple action/reaction. The Co-GENESIS feed-back loop of adaptation and variation informed life and natural selection advanced the most viable: Nematodes to Neanderthals. Evolution was at the speed of birth until the next giving-of-birth (passing on updated genes).

Nature *shaped* and *informed* all Pre-Human life.

100,000 years ago Humans evolved deliberative-consciousness—a binary state of perception in which we observe ourselves within the external world. When Humans etched complex patterns on an ostrich egg shell, they confirmed a conscious mindset—seeing, interpreting, and then projecting: painting, sculpture, and modifications into the world. Caves were replaced with hilltop fortifications and, with projectile weapons, Humans rose to dominate the farthest reaches of the Earth.

Humans *shape* and *inform* Nature, in their image and to their ends.

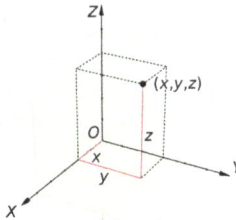

Figure 79. The Object as Defined by Cartesian Geometry

An object can be defined using the concepts of René Descartes, a French mathematician—the location of a point in 3-dimensional space (see above). In 3-D animation such as James Cameron's *Avatar*, all Cartesian points on a 3-D form are located via a two factor (binary) code processed by a computer, just as the RNA MIND in Co-GENESIS processes the binary code, The Code of Life, when plotting the points to form a baby's hand.

Human consciousness is evolution within the Mind. Humankind has essentially become self-evolving and, until now, we had not touched the biological foundation of our consciousness. Today, however, we stand at the precipice of destroying the masterwork we simply don't understand.

GLOSSARY

Adaptive characteristics – the traits of an individual that increase the chances of survival and reproduction in an environment.

AIA – American Institute of Architects, Washington. DC.

Algorithm – a set of instructions for accomplishing a task or a solving a problem.

Ancestral RNA Mind – RNA imprinted parental genes before conception of offspring: RNA(m) male legacy and RNA(f) female legacy.

Anticipatory RNA Mind – the half-cell sperm and half-cell egg imprinted genes which receive adaptive information during the parent's lifetimes to update the Applied RNA Mind of offspring.

Applied RNA Mind – the inherited RNA imprinted genes (male legacy & female legacy), updated by parental adaptive information: miRNA(m) and miRNA(f) in offspring.

Assimilation – a process by which an inherited adaptive trait that continues to enhance survival and reproduction in the future, becomes increasingly permanent. However, should the environmental circumstance which gave rise to the trait disappear in the future, the trait will fade away.

Binary – a binary is defined as "composed of or relating to two things." A binary is also a numerical system or *Code* that uses a two-symbol system.

The binary relationship of the RNA Mind (imprinted genes) is necessary to achieve homeostasis, *the fine tuning of bodily systems to set points. For example, you cannot control a car with an accelerator only, you must have the binary of independent braking and acceleration to maintain a set speed within a dynamic environment.*

Cellular death – the removal of body cells during growth or development which is directed by RNA through the process of *apoptosis*

Cellular division – the addition of body cells occurs via a DNA replication process, but a "start, and start here" instruction from RNA is required.

Circadian Rhythm – the 24-hour day/night cycle of the earth's rotation which informs the RNA biological timing mechanisms for optimum bodily function including: hormone levels, metabolism, perception, blood pressure, energy levels, sleep/wake, REM sleep and memory consolidation.

Circannular Rhythm – the annual rotation of the earth around the sun on its tilted axis generates the pattern of the seasons which informs a secondary timing mechanism for optimum bodily function, such as increasing biological resistance during an annual cold and flu season, or providing a second timing check for developmental stages and aging.

Code of Life – the instructions for making, regulating, and reproducing a living thing, embodied as the DNA helix in the nucleus of every cell at the start.

The DNA helix is the inactive form of information, a reference library from which RNA transcribes active RNA molecules. The DNA helix exists within an ocean of active RNA which controls its replication, modification, and destruction. ±80% of the helix becomes active RNA coding (genetic/ body) or non-coding (epigenetic/mind).

Co-evolution – the influence of closely associated animal species in their evolution via natural selection: advantageous traits pass from one generation to the next.

In humans, co-evolution occurs primarily via deliberative consciousness and language in real time, effectively an acceleration of evolution independent of the inheritance of traits via natural selection.

Co-GENESIS – *the Theory of RNA Primacy, the RNA Mind, Perception and Consciousness*

CRISPER Cas-9 – a genetic engineering technique to modify the genetic and/ or epigenetic material within the genome of living things.

Dark Matter – the approximately 75% of the human genome's material that is non-coding, epigenetic, not a regulatory function for the body, and thought by many to be non-functional. Most of these short sequences are transposons and pseudogenes, or repetitive copies of near-identical sequences, and fragments of viral origin.

The RNA Mind has evolved to a near-infinite spectrum of variation. The hyper-complexity of achieving human consciousness while maintaining a resilient social pattern of mindsets and passing down adaptive characteristics from generation-to-generation requires the full range of repetitive sequences within Dark Matter—the Master Toolbox of the RNA MIND— casting and recasting the acquired and retained array of variation potentials for over a billion years.

Deliberative consciousness – 1. *Deliberative* – relating to or intended for consideration or discussion. 2. *Consciousness* – the fact of awareness by the mind of itself and the world.

By definition a discussion requires two minds and in humans, the two voices in discussion are the left hemisphere maleness and the right hemisphere femaleness—our silent inner voice. Awareness of the mind of itself (internal), and observation of the world (external) are made possible by this deliberation which establishes a binary consensus of "here" vs. the objects being observed, "there." Deliberative consciousness is unique to humans.

DNA – is the organic molecule that stores information organized in the geometric form of the double helix.

DNA is the organic molecule that stores genetic and epigenetic information, the Code of Life, in the inactive form of the double helix. DNA is accessible by surrounding RNA which transcribes RNA molecules from the helix—active RNA forms of the genetic and epigenetic information.

Egg – the female half-cell genome, or gamete.

Embryogenesis – the process of initiation and development of an embryo from a zygote.

Fallopian Tube – two bilateral female structures that transport the ova (egg), from the ovary to the uterus (womb) each month.

Femaleness – the transgenerational legacy of female perception/behavior which is independent of anatomical sex (e.g., empathy, emotional intelligence, creativity, future-directed thinking). Femaleness is also referred to as the Female Legacy.

Fertilization – the fusion of male and female gametes to form a zygote.

Fertilization-to-Implantation – after fertilization, the single-cell zygote travels through the fallopian tube, completing a series of cellular divisions before entering the womb as a grouping of over 100 cells for implantation in the uterine lining (traverse varies: 6-12 days).

Gametes – a half-cell sperm or half-cell egg which form the zygote.

Gametogenesis – the process, approximately four months after fertilization, in which the gametes for the developing fetus are formed.

Gametogenesis passes down the updated ancestral male legacy RNA(m) to the sperm or, the updated ancestral female legacy RNA(f) to the egg.

Genes – the basic unit of heredity passed from a parent to offspring.

Co·GENESIS Working Assumption on Genes:

<u>RNA Binary</u>
(Imprinted) non-coding/epigenetic = 400

<u>RNA Mixed</u>
coding/genetic = 20,000
non-coding epigenetic = 20,000

Genetics – biological studies in heredity, particularly the mechanisms of hereditary transmission and the variation of inherited characteristics.

Genome – the complete set of genes or genetic material present in a cell or organism.

Genome – the complete set of genes which can be binary (non-coding/epigenetic) or unary (coding/genetic or non-coding/epigenetic) plus the sub-components and scripts of hereditary material utilized in the detailed variation of the Code of Life.

Genome editing – a method for making specific changes to the DNA of a cell or organism. It can be used to add, remove or alter DNA in the genome.

A method for making changes to the Code of Life when in the inactive form of the DNA helix in a cell or organism. If the cell is a body (somatic) cell the modification should stay within the body, however if the cell is a reproductive cell—a germline cell—or an embryo, the changes can pass on to future generations.

Genomic imprinting – See RNA Imprinted Genes.

Germline – the sex cells (eggs and sperm) that sexually reproducing organisms use to pass on their genomes from one generation to the next (parents to offspring). Egg and sperm cells are called germ cells, in contrast to the other cells of the body, which are called somatic cells.

Gestation – the period of development inside the womb between fertilization and birth.

Helix – the shape taken by the DNA molecule. A helix is a three-dimensional spiral, like the shape of a spring or the railing on a spiral staircase. A DNA molecule consists of two helixes intertwined.

Homeostasis – a state of balance among all the internal body systems needed for an organism to survive and function correctly.

Hypothalamus – a region of the brain located at the mid-brain centerline near the pituitary gland that plays a crucial role in many important functions, including: perception/behavior and homeostasis via hormonal and electrical signaling, maintaining daily physiological cycles.

Hypothalamus – the binary seat of biological intelligence and the RNA MIND, uniquely expressed in each individual—world view (personality).

Inner Voice or Silent Voice – the subjective experience of language in the absence of overt and audible articulation.

Instinctive/subconscious – the inherent disposition of a living organism toward a particular behavior.

lncRNA (long non-coding RNA) – are RNA imprinted molecules larger than 200 nucleotides.

Maleness – the transgeneration legacy of male perception/behavior which is independent of anatomical sex (e.g. decisive, assertive, analytical, low empathy, impatient). Maleness is also referred to as the Male Legacy.

Mediate – accomplishing tasks indirectly with the aid of an intervening medium.

Mind/Gender – a personal conception of oneself as male or female (or rarely, both or neither). This concept is intimately related to the concept of gender role, which is defined as the outward manifestations of personality that reflect the gender identity.

miRNA (micro-RNA) – a class of non-coding RNAs that play essential roles in regulating gene expression and communicating between cells.

Mitochondria – small organelles within the cell that have their own RNA and DNA and convert food energy to the energy used by the cell in the form of the ATP molecules.

Non-Mendelian or Non-Crick/Watson – biological phenomena that are inconsistent with the DNA-centric conceptual framework of evolutionary biology—the Modern Synthesis or Neo-Darwinian Theory.

Nucleotides – DNA/RNA building blocks of information.

Pangenesis – a hypothetical mechanism of heredity proposed by Charles Darwin in which *gemmules* (with adaptive information) are passed from body cells to the bloodstream, to the gonads, and then onto the reproductive cells.

Parental – the preceding generation from which hybrid offspring are produced.

Perception/Behavior – the processing of sensory stimuli into a given pattern of response.

Plasticity – the capacity of organisms or cells to alter their phenotype (traits) in response to changes in their environment.

Predisposition – the state of being likely to behave in a certain way.

RNA – ribonucleic acid (RNA) is a molecule consisting of a long linear chain of nucleotides.

The organic molecule that is the active form of heritable information stored in the DNA helix.

RNA Binary Mind – the immortal binary of male/female differentness joined in each birth as the Mind and passed on in the conception of offspring via the male half-cell and female half-cell gametes (the next generational Mind).

RNA expression – the process by which a gene within a cell is instructed to make RNA or proteins.

RNA Imprinted Genes – the process by which only one copy of a gene in an individual (either from their mother or their father) is expressed, while the other copy is suppressed.

RNA processing – the term collectively used to describe the sequence of events through which RNA matures a transcription from the helix to an RNA or a protein.

RNA World – the RNA world hypothesis that suggests that life on Earth began with a simple RNA molecule that could copy itself without help from other molecules.

Silencing – generally used to describe the "switching off" of male or female parental (ancestral) genes by genomic imprinting.

Sperm – the male sex cell, or gamete.

Synthesis – the process of combining two or more components to produce an entity.

Thermal balance – occurs when the sum of all the different types of heat flow into and out of a body is in balance at a given set point—it can be said to be in equilibrium (i.e. 98.6° F in Humans).

Transcription – the process by which an RNA molecule extracts information from the DNA helix. This RNA copy carries the information needed to make proteins in a cell or it may be an independent RNA molecule.

The process by which RNA makes an RNA active version of DNA inactive information.

Transgenerational – the transmission of RNA epigenetic markers from one organism to the next (i.e., from parent to child) that affects the traits of offspring without altering the primary structure of DNA.

UIA – International Union of Architects, Paris, France.

Ultra-conserved element – genome segment longer than 200 base pairs that is absolutely conserved, with no insertions or deletions and 100% identity between humans, mouse, and rat genomes (common ancestor).

Uterus – the main hormone-responsive, secondary sex organ of the female reproductive system in humans, and most other mammals, typically referred to as the womb.

Vesicles – cellular envelopes that are used to transport materials or liquids from one place to another.

YBP – A year designation alternative to the widely-used but Christian-oriented designations BC and AD and their secular equivalents BCE and CE. YBP stands for years before present.

Zygote – fertilized egg cell that results from the union of a female gamete (egg, or ovum) with a male gamete (sperm).

ACKNOWLEDGEMENTS

A chance meeting at the Harvard Faculty Club in 2005 with Edward O. Wilson led to a discussion of his seminal book *Consilience* (1998), and prompted a conceptual diagram I penned later that day: "2 Sides of Consciousness." This was the beginning of an eighteen-year quest to understand the binary basis of Human consciousness and perception. Six years later, focused on an on-going architectural practice, I had advanced no further than a thirty-page treatment. However, in 2011 our firm won the commission for the Ecuador Genomic Biology and Biodiversity Institute in Quito and our client, Manuel Balderon, Head of the Secretariate of Science and Technology, was set to host the first International Congress on Biodiversity. Two champions of biodiversity were to be headliners: Edward O. Wilson and Peter H. Raven, President Emeritus of Missouri Botanical Garden—an early client of our firm. Participating with Wilson, Raven and leadership of the Congress in May of 2011 was the inspiration of a lifetime. With new momentum, I completed the book *A Convergence of Two Minds* in 2015.

Two Minds proved to be foundational to the assertion of RNA primacy in *Co·GENESIS*—a proof of concept based on subsequent evidence from the leading edge of research in biology and the natural sciences. David Kenneth Specter, an architect and close friend, introduced me to a sage advisor following Quito: Donald S. Lamm, previously Chairman of W. W. Norton in New York City before his active retirement to Santa Fe and the Board of the School for Advanced Research (SAR). Don's advice, after seeing my timidity and lengthy referrals to other scientific opinions was straight-forward—footnote sources for the back-of-the-book, but find your own voice in the body of the work. Janet Adams Strong, a distinguished editor, persistently guided me to an accessible telling of complex and, as it turned out, revolutionary assertions. Reviews of the manuscript were by Glen Gehar, Director of Evolutionary Studies and Chair of Psychology, State University of New York at New Paltz and Donald Watson, Author and Dean Emeritus of the RPI School of Architecture. Donald suggested the "Chronology of an Idea" and the Glossary as dual worldviews—RNA-and-DNA in Co·GENESIS. Informal manuscript-in-progress reviews were provided by Peter H. Raven, President Emeritus of Missouri Botanical Garden, Carl N. McDaniel, Professor Emeritus RPI/Biology, Miriam Feuerman, Faculty of Cell Biology, NY Downstate Health Sciences University, Robert Koester AIA, Professor of Architecture at Ball State University, and Coke Anne Wilcox, President of the Board of Trustees of Pequot Library, Southport, Connecticut and her husband, the artist Jarvis Wilcox, who went beyond their insightful reviews to sponsor a book-signing event for *A Convergence of Two Minds*.

NOTES

FOREWORD

1) DNA (inactive code) consists of molecular strands of heritable information in helical form that stay within the cell nucleus, referred to as *nucleated DNA*.

2) MM Patten, L Ross, JP Curley, DC Queller, R Bonduriansky and JB Wolf: The evolution of genomic imprinting: theories, predictions and empirical tests. *Heredity* (2014) **113**, 119–128; published online 23 April 2014

3) Cornelia Meinert, Iuliia Myrgorodska, Pierre de Marcellus, Thomas Buhse, Laurent Nahon, et al., Ribose and related sugars from ultraviolet irradiation of interstellar ice analogues. *SCIENCE*, (2016) doi:10.1126/science.aad8137.

4) Biswajit Samanta and Gerald J. Joyce. A reverse transcriptase ribozyme *eLife* (2017) **6**: e31153. DOI:10.7554/eLife.31153

5) JD Perez, ND Rubinstein, C Dulac, New Perspectives on Genomic Imprinting, an Essential and Multifaceted Mode of Epigenetic Control in the Developing and Adult Brain. *Annual Review of Neuroscience* (2016) **8**; 39:347-84. DOI: 10.1146/annurev-neuro-061010-113708.

6) Judith Reichmann, Bianca Nijmeijer, M. Julius Hossain, Manuel Eguren, Isabell Schneider, Antonio Z. Politi, M. Julia Roberti, Lars Hufnagel, Takashi Hiiragi, Jan Ellenberg. Dual-spindle formation in zygotes keeps parental genomes apart in early mammalian embryos. *SCIENCE* **361**, 189–193 (2018)

7) Anna Lorenc, Miriam Linnenbrink, Inka Montero, Markus B. Schilhabel, and Diethard Tautz. Genetic Differentiation of Hypothalamus Parentally Biased Transcripts in Populations of the House Mouse Implicate the Prader–Willi Syndrome Imprinted Region as a Possible Source of Behavioral Divergence, *Molecular Biology and Evolution* (2014) **31**(12):3240–3249 doi:10.1093/molbev/msu257

8) *Darwin Correspondence Project/Cambridge University Library,* darwinproject.ac.uk Charles Darwin to J. D. Hooker: 23 February [1868].

9) David Cyranoski, What CRISPR-baby prison sentences mean for research; Chinese court sends strong signal by punishing He Jiankui and two colleagues. *Nature* **577**, 154-155 (2020) doi:*10.1038/d41586-020-00001-y*

10) Isaacson, Walter: *The Code Breaker: Jennifer Doudna, Gene Editing, and the Future of the Human Race*, Simon and Schuster, New York, NY (2021).

CHAPTER ONE: THE BRIDGE

1) *Ecologically-Informed Design* was the term Croxton Collaborative Architects used to describe their approach to Architecture (reconciling built and natural environments). David Gottlieb and Mike Italiano, coordinating with Croxton during the pre-formation period of the US Green Building Council (USGBC), successfully argued that *Green Building* would be more memorable. Co·GENESIS informs even deeper levels of design—a humanistic or *human-centered* design.

2) *Buildings One* | February 4, 2016, HVAC Outdoor Air Ventilation Standard ASHRAE Standard 62.1- 2013— "By example, the HVAC system, in a building built in 1985, could provide a maximum of 7 cubic feet per minute (CFM) per person outdoor air, while the current ASHRAE 62.1-2013 Standard requires 17 CFM per person outdoor air."

3) Jason E. McDermott, Christopher S. Oehmen, Lee Ann McCue, Eric Hill, Daniel M. Choi, Jana Stöckel, Michelle Liberton, Himadri B. Pakrasic, and Louis A. Shermand. A model of cyclic transcriptomic behavior in the cyanobacterium Cyanothece sp. ATCC 51142. *Molecular BioSystems* Vol. 7, 2407–2418 (2011) DOI: 10.1039/c1mb05006k

4) Bae S-A, Fang MZ, Rustgi V, Zarbl H and Androulakis IP. At the Interface of Lifestyle, Behavior, and Circadian Rhythms: Metabolic Implications. *Frontiers of Nutrition* (2019) **6**:132. DOI: 10.3389/fnut.2019.00132

5) XC Dopico, Marina Evangelou, Ricardo Ferreira, H Guo, Marcin Pekalski, Deborah Smyth, Nicholas Cooper, Oliver Burren, Tony Fulford, Branwen Hennig, Andrew Mulberg, AG Ziegler, E Bonifacio, C Wallace, JA Todd. Widespread seasonal gene expression reveals annual differences in human immunity and physiology. *Nature Communications* Vol. **6**. (May 2015)

6) Annemarie Dosen and Michael Ostwald. Prospect and refuge theory: Constructing a critical definition for architecture and design. International *Journal of Design in Society* (2013) **6**: 9-23. 10.18848/2325-1328/CGP/v06i01/38559.

CHAPTER TWO: THE BASELINE

1) Cornelia Meinert, Iuliia Myrgorodska, Pierre de Marcellus, Thomas Buhse, Laurent Nahon, Søren V. Hoffmann, Louis Le Sergeant d'Hendecourt, Uwe J. Meierhenrich, Ribose and related sugars from ultraviolet irradiation of interstellar ice analogs SCIENCE **352**, 6282, pp. 208-212 (2016). DOI: 10.1126/science.aad8137

2) Ben K. D. Pearce, Ralph E. Pudritz, Dmitry A. Semenov, and Thomas K. Henning. Origin of the RNA world: The fate of nucleobases in warm little ponds. *PNAS* (2017) **114** (43) 11327-11332

3) Sidney Becker, Jonas Feldmann, Stefan Wiedemann, Hidenori Okamura, Christina Schneider, Katharina Iwan, Antony Crisp, Martin Rossa, Tynchtyk Amatov, Thomas Carell. Unified prebiotically plausible synthesis of pyrimidine and purine RNA ribonucleotides. *SCIENCE* (2019) **366**, 6461 pp. 76-82 DOI: 10.1126/science.aax2747

4) Biswajit Samanta and Gerald J. Joyce, A reverse transcriptase ribozyme *eLife* (2017) **6**: e31153 DOI: 10.7554/eLife.31153

CHAPTER THREE: THE BINARY

1) MM Patten, L Ross, JP Curley, DC Queller, R Bonduriansky and JB Wolf: The evolution of genomic imprinting: theories, predictions and empirical tests. *Heredity* (2014) **113**, 119–128; published online 23 April 2014.

2) Ågren JA, Clark AG (2018) Selfish genetic elements. PLoS Genetics **14** (11): e1007700. doi.org/10.1371/journal.pgen.1007700

3) Serena Tucci, Samuel H. Vohr, Rajiv C. Mccoy, Benjamin Vernot, Matthew R. Robinson, Chiara Barbieri, Brad J. Nelson, Wenqing Fu, Gludhug A. Purnomo, Herawati Sudoyo, Evan E. Eichler, Guido Barbujani, Peter M. Visscher, Joshua M. Akey, Richard E. Green.
Evolutionary history and adaptation of a human pygmy population of Flores Island, Indonesia. *SCIENCE* (2018) vol. **361**, 6401: 511-516. doi:10.1126/science.aar8486

4) Iain McGilchrist. *The MASTER and his EMISSARY: The Divided Brain and the Making of the Modern World*. Yale Univesity Press, New Haven (2009).

5) Paul Ehrlich. *Human Natures: Genes, Cultures, and the Human Prospect*. Island Press, Washington, DC (2000).

CHAPTER FOUR

1) Daniela Sammler, Marie-He´ le` ne Grosbras, Alfred Anwander, Patricia E.G. Bestelmeyer, Pascal Belin. Dorsal and Ventral Pathways for Prosody. *Current Biology* **25**, 3079–3085 (2015) dx.doi.org/10.1016/j.cub.2015.10.009

2) Mercedes Conde-Valverde, Ignacio Martínez, Rolf M. Quam, Manuel Rosa, Velez Alex D., Carlos Lorenzo, Pilar Jarabo, José María Bermúdez De Castro, Eudald Carbonell, Juan Luis Arsuaga.
Neanderthals and Homo sapiens had similar auditory and speech capacities. *Nature Ecology and Evolution* **5**, 609–615 (2021). doi.org/10.1038/s41559-021-01391-6

3) Gokhman D, Nissim-Rafinia M, Agranat-Tamir L, Housman G, García-Pérez R, Lizano E, Cheronet O, Mallick S, Nieves-Colón MA, Li H, Alpaslan-Roodenberg S, Novak M, Gu H, Osinski JM, Ferrando-Bernal M, Gelabert P, Lipende I, Mjungu D, Kondova I, Bontrop R, Kullmer O, Weber G, Shahar T, Dvir-Ginzberg M, Faerman M, Quillen EE, Meissner A, Lahav Y, Kandel L, Liebergall M, Prada ME, Vidal JM, Gronostajski RM, Stone AC, Yakir B, Lalueza-Fox C, Pinhasi R, Reich D, Marques-Bonet T, Meshorer E, Carmel L. Differential DNA methylation of vocal and facial anatomy genes in modern humans. *Nature Communications.* **11** (1):1189 (2020). doi: 10.1038/s41467-020-15020-6.

4) Julia Galway-Witham and Chris Stringer. How did Homo sapiens evolve? *SCIENCE* (2018) Vol **360**, 6395, pp. 1296-1298, DOI: 10.1126/science.aat6659

5) Mark Lipson, Isabelle Ribot, Swapan Mallick, Nadin Rohland, Iñigo Olalde, Nicole Adamski, Nasreen Broomandkhoshbacht, Ann Marie Lawson, Saioa López, Jonas Oppenheimer, Kristin Stewardson, Raymond Neba'ane Asombang, Hervé Bocherens, Neil Bradman, Brendan J. Culleton, Els Cornelissen, Isabelle Crevecoeur, Pierre de Maret, Forka Leypey, Mathew Fomine, Philippe Lavachery, Christophe Mbida Mindzie, Rosine Orban, Elizabeth Sawchuk, Patrick Semal, Mark G. Thomas, Wim Van Neer, Krishna R. Veeramah, Douglas J. Kennett, Nick Patterson, Garrett Hellenthal, Carles Lalueza-Fox, Scott MacEachern, Mary E. Prendergast & David Reich. Ancient West African foragers in the context of African population history. *NATURE* **577**, pages 665–670 (2020)

6) Sriram Sankararaman, Swapan Mallick, Michael Dannemann, Kay Prüfer, Janet Kelso, Svante Pääbo, Nick Patterson, David Reich. The genomic landscape of Neanderthal ancestry in present-day humans. *NATURE* **507**, pp. 354-357 (2014) doi.org/10.1038/nature12961

By examining the genomes of over a thousand present-day Humans to find genomic regions that were high or low in the small amount of Neanderthal heritable material we still possess, researchers were able to identify areas that were highly resistant to the interbreeding of Humans and Neanderthals. For example, the consistent indication on the modern-day Human X chromosome of very low Neanderthal ancestry is convincing evidence of *male hybrid sterility* in the offspring of any such interbreeding. A similar effect is observed when a horse and a donkey breed across species to create a mule. *The male mule is born sterile and cannot reproduce.*

An additional and stunning assertion, germane to Co·GENESIS, was made:

Hybrid sterility is not the only factor responsible for selection against Neanderthal material as Neanderthal ancestry is also depleted in conserved pathways such as RNA processing.

7) John J. Shea and Matthew L. Sisk. Complex Projectile Technology and *Homo sapiens* Dispersal into Western Eurasia. *Paleo Anthropology*, 2010: 100–122. doi:10.4207/PA.2010.ART36.

8) Sahle Y, Brooks AS. Assessment of complex projectiles in the early Late Pleistocene at Aduma River, Ethiopia. *PLoS ONE* (2019) **14** (5): e0216716. doi.org/10.1371/journal.pone.0216716.

9) Pearce, Stringer and Dunbar. "New insights into differences in brain organization between Neanderthals and anatomically modern humans," the Proceedings of the Royal Society of London. Series B, Biological sciences, (2013) doi: 10.1098/rspb.2013.0168.

10) Sankararaman et al, Genomic Landscape, p.56.

11) CS Henshilwood, F d'Errico, KL van Niekerk, L Dayet, A Queffelec, Pollarolo L. An abstract drawing from the 73,000-year-old levels at Blombos Cave, South Africa. *NATURE* **562** (7725): pp.115-118. (2018) DOI: 10.1038/s41586-018-0514-3.

12) Pierre-JeanTexier, Guillaume Porraz, John Parkington, Jean-Philippe Rigaud, Cedric Poggenpoel,*Christopher Miller, Chantal Tribolo, Caroline Cartwright, Aude Coudenneau, Richard Klein, Teresa Steele, Christine Verna. A Howiesons Poort tradition of engraving ostrich eggshell containers dated to 60,000 years ago at Diepkloof Rock Shelter, South Africa. *PNAS* (2010) **107** (14) 6180-6185; DOI: 10.1073/pnas.0913047107.

13) DE Rosso, Martí A Pitarch, F d'Errico. Middle Stone Age Ochre Processing and Behavioral Complexity in the Horn of Africa: Evidence from Porc-Epic Cave, Dire Dawa, Ethiopia. *PLoS ONE* (2016) **11** (11): e0164793. doi.org/10.1371/journal.pone.0164793.

CHAPTER FIVE: THE MECHANISMS

1) Judith Reichmann, Bianca Nijmeijer, M. Julius Hossain, ManuelEguren, Isabell Schneider, Antonio Z. Politi, M. Julia Roberti, Lars Hufnagel, Takashi Hiiragi, Jan Ellenberg. Dual-spindle formation in zygotes keeps parental genomes apart in early mammalian embryos. *SCIENCE* (2018) Vol **361**, 6398 189-193 DOI: 10.1126/science. aar7462

2) Zielinska AP, Schuh M. Double trouble at the beginning of life. *SCIENCE.* **13**; 361(6398):128-129. doi: 10.1126/science.aau3216.

3) Darwin, Charles. *The Variation of Animals and Plants under Domestication.* London: John Murray, 1868 (two editions, 1868 and 1875). www.biodiversitylibrary.org/bibliography/61215#/ summary.

4) O'Brien J, Hayder H, Zayed Y, and Peng C (2018) Overview of MicroRNA Biogenesis, Mechanisms of Actions, and Circulation. *Frontiers in Endocrinology* **9**: 402. doi: 10.3389/fendo.2018.00402

5) Upasna Sharma, Fengyun Sun, Colin C. Conine, Brian Reichholf, Shweta Kukreja, Veronika A. Herzog, Stefan L. Ameres, and Oliver J. Rando. Small RNAs Are Trafficked from the Epididymis to Developing Mammalian Sperm. *Developmental Cell* **46**, pp. 481–494, (2018) doi.org/10.1016/j.devcel.2018.06.023

6) Nicole Gross, Jenna Kropp, and Hasan Khatib. MicroRNA Signaling in Embryo Development. *Biology* **6**, 34; (2017) doi:10.3390/biology6030034

7) Alessandro Fatica and Irene Bozzoni. Long non-coding RNAs: new players in cell differentiation and development. *NATURE* **15**: 7-21(2014) www.nature.com/reviews/genetics

8) Karen P. Maruska and Russell D. Fernald. Social Regulation of Gene Expression in the Hypothalamic-Pituitary-Gonadal Axis. PHYSIOLOGY **26**: pp. 412–423, (2011) doi:10.1152/physiol.00032.2011

Reproduction is a critically important event in every animal's life and in all vertebrates is controlled by the brain via the hypothalamic-pituitary-gonadal (HPG) axis. In many species, this axis, and hence reproductive fitness, can be profoundly influenced by the social environment. Here, we review how the *reception of information in a social context causes genomic changes at each level of the HPG axis.*

9) Chunyu Cao, Yifei Ding, Xiangjun Kong, Guangde Feng, Wei Xiang, Long Chen, Fang Yang, Ke Zhang, Mingxing Chu, Pingqing Wang, Baoyun Zhang. Reproductive role of miRNA in the hypothalamic-pituitary axis. *Molecular and Cellular Neuroscience* **88**, pp. 130-137 (2018) doi. org/10.1016/j.mcn.2018.01.008

CHAPTER SIX: TRANSGENERATIONAL PATHWAYS

1) Nishikawa, K., Kinjo, A.R. Mechanism of evolution by genetic assimilation. *Biophysical Reviews* **10**, 667–676 (2018). doi.org/10.1007/s12551-018-0403-x

2) Olivia Ho-Shing, Catherine Dulac, Influences of genomic imprinting on brain function and behavior, *Current Opinion in Behavioral Sciences*, **25**, Pages 66-76, (2019) doi.org/10.1016/j.cobeha.2018.08.008.

3) Daniela Sammler, Marie-He′ le` ne Grosbras, Alfred Anwander, Patricia E.G. Bestelmeyer, Pascal Belin. Dorsal and Ventral Pathways for Prosody. *Current Biology* **25**, 3079–3085 (2015) dx.doi.org/10.1016/j.cub.2015.10.009

4) Elissa D. Pastuzyn, Cameron E. Day, Rachel B. Kearns, Madeleine Kyrke-Smith, Andrew V. Taibi, John McCormick, Nathan Yoder, David M. Belnap, Simon Erlendsson, Dustin R. Morado, John A.G. Briggs, Ce′dric Feschotte, and Jason D. Shepherd. The Neuronal Gene Arc Encodes a Repurposed

Retrotransposon Gag Protein that Mediates Intercellular RNA Transfer. *Cell* **172**, 275–288, (2018)
doi: 10.1016/j.cell.2017.12.024.

5) Izawa S, Chowdhury S, Miyazaki T, Mukai Y, Ono D, Inoue R, Ohmura Y, Mizoguchi H, Kimura K, Yoshioka M, Terao A, Kilduff TS, Yamanaka A. REM sleep-active MCH neurons are involved in forgetting hippocampus-dependent memories. *Science* **365** (6459):1308-1313 (2019)
doi: 10.1126/science. aax 9238.

6) Houle D, Bolstad GH, van der Linde K, Hansen TF. Mutation predicts 40 million years of fly wing evolution. *Nature* (2017) **548** (7668):447-450.
doi: 10.1038/nature23473.

7) Matsuura, Kenji. "Genomic imprinting and evolution of insect societies." *Population Ecology* **62.1** (2020): 38-52.
doi.org/10.1002/1438-390X.12026

8) Linneweber GA, Andriatsilavo M, Dutta SB, Bengochea M, Hellbruegge L, Liu G, Ejsmont RK, Straw AD, Wernet M, Hiesinger PR, Hassan BA. A neurodevelopmental origin of behavioral individuality in the Drosophila visual system. *Science* **6**; 367 6482) 1112-1119. (2020)
doi: 10.1126/science. aaw 7182.

CHAPTER SEVEN: SIGNIFICANCE & MEANING

1) Mitochondria. Small organelles within the cell that have their own RNA and DNA and convert food energy to the energy used by the cell: Atps.

2) Wilcox, Christie. Why Sex? Biologists Find New Explanations. *Quanta Magazine.* (April 2020) www.quantamagazine.org/why-sex-biologists-find-new-explanations-20200423/.

3) NIH: National Human Genome Research Institute. International Human *Genome Sequencing Consortium Publishes Sequence and Analysis of the Human Genome.* February 12, 2001. www.genome.gov/10002192/2001-release-first-analysis-of-human-genome.

4) Choi Sang-Han, Jeong Gangwon, Hwang Young-Eun, Kim Yong-Bo, Lee Haigun, Cho Zang-Hee. Track-Density Ratio Mapping with Fiber Types in the Cerebral Cortex Using Diffusion-Weighted MRI. *Frontiers in Neuroanatomy* **15**, (2021) doi.10.3389/fnana.2021.715571

5) NIH: Human Connectome Project (2011-2014) www.humanconnectomeproject.org/category/news/

6) Brainnetome Center, National Laboratory of Pattern Recognition, Institute of Automation, The Chinese Academy of Sciences, 100190, PR China www.brainnetome.org

7) NIH: Human Connectome Project (2014)

8) Wilson, Edward O., Consilience: *The Unity of Knowledge*, Alfred A. Knopf, New York, N. Y., 1998

9) Cobb, M (2017) 60 years ago, Francis Crick changed the logic of biology. *PLoS Biology* **15** (9): e2003243. https://doi.org/10.1371/journal.pbio.2003243

10) Ibid., p. 126.

11) Cyranoski, David. What CRISPR-baby prison sentences mean for research: Chinese court sends strong signal by punishing He Jiankui and two colleagues. *Nature* **577**, 154-155 (2020) doi.org/10.1038/d41586-020-00001-y

12) Gorvett, Zaria. The genetic mistakes that could shape our species. *BBC Future* (12 April 2021). www.bbc.com/future/article/20210412-the-genetic-mistakes-that-could-shape-our-species.

13) Isaacson, Walter: *The Code Breaker: Jennifer Doudna, Gene Editing, and the Future of the Human Race*, Simon and Shuster, New York, NY (2021).

14) Jennifer A. Doudna and Samuel H. Sternberg. *A Crack in Creation: Gene Editing and the Unthinkable Power to Control Evolution*. Houghton Mifflin Harcourt, New York (2017)

15) Greely, Henry T. *CRISPR People: The Science and Ethics of Editing Humans*. MIT Press, Cambridge, MA 2021

16) Jan Zalasiewicz, Colin N. Waters, Mark Williams, Anthony D. Barnosky, Alejandro Cearreta, Paul Crutzen, Erle Ellis, Michael A. Ellis, Ian J. Fairchild, Jacques Grinevald, Peter K. Haff, Irka Hajdas, Reinhold Leinfelder, John McNeill, Eric O. Odada, Clément Poirier, Daniel Richter, Will Steffen, Colin Summerhayes, James P.M. Syvitski, Davor Vidas, Michael Wagreich, Scott L. Wing, Alexander P. Wolfe, Zhisheng An, Naomi Oreskes. When did the Anthropocene begin? A mid-twentieth century boundary level is stratigraphically optimal. *Quaternary International*, **383**, 196-203, (2015) doi.org/10.1016/j.quaint.2014.11.045.

17) Suzana Herculano-Houzel. The remarkable, yet not extraordinary, human brain as a scaled-up primate brain and its associated cost. *PNAS* **109** 10661-10668 doi.org/10.1073/pnas.1201895109

18) Manolis Kellis, Barbara Wold, Michael P. Snyder, Bradley E. Bernstein, Anshul Kundaje, Georgi K. Marinov, Lucas D. Ward, Ewan Birney, Gregory E. Crawford, Job Dekker, Ian Dunham, Laura L. Elnitski, Peggy J. Farnham, Elise A. Feingold, Mark Gerstein, Morgan C. Giddings, David M. Gilbert, Thomas R. Gingeras, Eric D. Green, Roderic Guigo, Tim Hubbard, Jim Kent, Jason D. Lieb, Richard M. Myers, Michael J. Pazin, Bing Ren, John A. Stamatoyannopoulos, Zhiping Weng, Kevin P. White, Ross C. Hardison. Defining functional DNA elements. *Proceedings of the National Academy of Sciences* (2014) **111** (17) 6131 – 6138, doi: 10.1073/pnas.1318948111

19) Ahituv N, Zhu Y, Visel A, Holt A, Afzal V, Pennacchio LA, et al. (2007) Deletion of Ultraconserved Elements Yields Viable Mice. *PLoS Biol* **5** (9): e234. doi.org/10.1371/journal.pbio.0050234

20) Snetkova V, Ypsilanti AR, Akiyama JA, Mannion BJ, Plajzer-Frick I, Novak CS, Harrington AN, Pham QT, Kato M, Zhu Y, Godoy J, Meky E, Hunter RD, Shi M, Kvon EZ, Afzal V, Tran S, Rubenstein JLR, Visel A, Pennacchio LA, Dickel DE. Ultraconserved enhancer function does not require perfect sequence conservation. *Nature Genetics.* **53** (4):521-528. (2021) doi: 10.1038/s41588-021-00812-3.

21) Hecker N, Hiller M. A genome alignment of 120 mammals highlight ultraconserved element variability and placenta-associated enhancers. *Gigascience.* 2020 Jan 1; **9** (1): giz159. doi: 10.1093/gigascience/giz159.

22) Saha, K., Sontheimer, E.J., Brooks, P.J. et al. The NIH Somatic Cell Genome Editing program. *Nature* **592**, 195–204 (2021). doi.org/10.1038/s41586-021-03191-1

23) Gregorio Alanis-Lobato, Jasmin Zohren, Afshan McCarthy, Norah M. E. Fogarty, Nada Kubikova, Emily Hardman, Maria Greco, Dagan Wells, James M. A. Turner, Kathy K. Niakan. Frequent loss of heterozygosity in CRISPR-Cas9–edited early human embryos. *Proceedings of the National Academy of Sciences.* **118** (22) e2004832117 doi: 10.1073/pnas.2004832117

24) Mattick, John S. et al. Long non-coding RNAs: definitions, functions, challenges, and recommendations. *nature reviews molecular cell biology* **24** (June 2023), 430-447. doi.org/10.1038/541580-022-00566-8

25) Greenberg, D. M., Decety, J., & Gordon, I. The social neuroscience of music: Understanding the social brain through human song. *American Psychologist*, **76** (7), 1172-1185. (2021). dx.doi.org/10.1037

IMAGE CREDITS

COVER

Front: Background: A remix from Henri Rousseau's *Virgin Forest with Sunset* (1910) in the collection of the Kunstmuseum Basel Museum (CCO 1.0) Universal Public Domain Dedication.

Front: Adam & Eve: Albrecht Dürer (1507), Getty Images (Adam: #463896821) (Eve: #463896817) Editorial License.

FOREWORD

Figure 1: Male | Female Legacy Separation
Microscope slide of zygote: Reichmann et al, *SCIENCE* (July 2018)
Artwork & Text by Author

Figure 2: Three Generations of RNA Body and RNA Mind
Artwork & Concept by Author

Figure 3: Mammalian Genomic Imprinting: Pathways of the RNA Mind
Artwork & Concept by Author

CHAPTER 1: THE BRIDGE

Figure 4: Circadian Rhythm: Day/Night Cycle and Biological Function
Interpretation, Graphics and Text by Author. Background Image from Salk Institute Lecture: Panda, Satchin, Circadian Rhythm During This Time of Social Distancing, June 4, 2020.
https://www.youtube.com/watch?v=11zPmbx9QF4

Figure 5: Architectural Integration: Solar Traverse & Work Environment
Illustration: Croxton Collaborative Architects, NRDC HQ, NYC

Figure 6: Cyanobacteria Cell: RNA Regulation of Genes Day-to-Night
Pacific Northwest National Lab (PNNL), Jason McDermott et al.
Molecular Bio-Systems **7**, **8**, 2011,
Overlay, Graphics and Text by Author.

Figure 7: Circadian Integration: Direct @ Circulation, Indirect @ Classroom
Illustration: Croxton Collaborative Architects, Rinker Hall, University of Florida, Gainsville, Florida.

Figure 8: Daylighting Retrofit: Light Slot, Ceiling Geometry & New Exterior
Illustration: Croxton Collaborative Architects, Wooster Hall, SUNY New Paltz, NY.

Figure 9:   Biological Alignment: Bodily Functions and Time of Day
Adaptation by Author for REM Sleep/Memory Reconciliation
Baseline image: The Road to Health Series: Light and Sleep.
Kilichowski, S., healthyplanbyann.com

Figure 10: Seasonal Variation of Beam Light: Main Stair Markers & South Porch
Illustration: Croxton Collaborative Architects, Wooster Hall,
SUNY New Paltz, NY.

Figure 11: Main Stair @ Summer Solstice: 8 Destinations Visible at Entry
Photo by Tim Hursley for Croxton Collaborative Architects, PC.,
Wooster Hall, SUNY New Paltz, NY.

Figure 12: Hypothalamus: Neurological/Endocrine Center of the RNA Mind
Image: 3-D anatomy of the brain, Rights purchased from Science
Photo Libary, London. Image #CO17/0470.
Artwork & Text by Author.

CHAPTER 2: THE BASELINE

Figure 13: Origin of the RNA World (Pre-biotic)
Author's Freehand Sketch (13), re-interpretation from Ben K. D.
Pearce et al., Origin of the RNA world: The fate of nucleobases in
warm little ponds. *PNAS* (2017) **114** (43) *Figure 2*, p. 11329.

Figure 14: Origin of the RNA World (Biotic).
Author's Freehand Sketch (14), interpretation from Ben K. D.
Pearce et al., Origin of the RNA world: The fate of nucleobases in
warm little ponds. *PNAS* (2017) **114** (43) *Figure 2*, p. 11329.

Figure 15: Transition from the RNA World
Adaptation by Author: Conceptual Transition: RNA-to-DNA
Baseline Images in *Microbiology*
Microbiology/OpenStax/American Society for Microbiology Press.
opentextbc.ca/microbiologyopenstax/chapter/
structure-and-function-of-rna/

Figure 16: Helix-to-RNA-to-tRNA Functional Molecule
Adaptation by Author of Baseline Images in *Microbiology*
Microbiology/OpenStax/American Society for Microbiology Press.
opentextbc.ca/microbiologyopenstax/chapter/
structure-and-function-of-rna/

Figure 17: Master RNA Organelle—RIBOSOME
Image under Creative Commons License: Unmodified.
Source: Fvoigtsh - Own work, CC BY-SA 3.0
File:80S 2XZM 4A17 4A19.png

Figure 18: **RIBOSOME rRNA Subunit: Over 1,500 Information Bars**
Secondary Mapping" of sub-unit of the Ribosome (rRNA)
Noller Lab: Center for Molecular Biology of RNA
Thermus thermophilus 16S rRNA Secondary Structure
rna.ucsc.edu/rnacenter/images/figs/thermus _16s_2ndry.jpg

CHAPTER 3: THE BINARY

Figure 19: **99% of GENES: Mixture of all Ancestral Genes**
Illustration by Author

Figure 20: **1% of GENES: Male Only and Female Only Ancestral Lines**
Illustration & Concept by Author

Figure 21: **Theories of GENOMIC IMPRINTING**
Composition and Text by Author
Upper Image: Ågren JA, Clark AG (2018) Selfish genetic
elements. *PLoS Genetics* **14** (11): e1007700.
Lower Image & Concept by Author

Figure 22: **GENOMIC IMPRINTING as a BINARY CODE**
Graphics, Text & Concept by Author
Morphology Image: by Author.
Regulatory Image: Thermoregulation in the human body, *Encyclo-
pedia of Occupational Health and Safety* (CISILO, 2013)
www.ukm.my/jkukm/wp-content/uploads/2021/3303/05.pdf
RNA Binary Image: by Author.

Figure 23: **The Male Binary RNA MIND**
Illustration & Concept by Author

Figure 24: **The Female Binary RNA MIND**
Illustration & Concept by Author

Figure 25: **PRE-HUMAN ADAPTATION of Perception/Behavior**
Illustration & Concept by Author

Figure 26: **NEUROLOGICAL PREDISPOSITION of the Hemispheres**
Illustration and Interpretation by Author
Original Concept: Iain McGilchrist. *The MASTER and his
EMISSARY: The Divided Brain and the Making of the Modern World.*
Yale University Press, New Haven (2009).

Figure 27: **A Spectrum of NEUROLOGICAL GENDERS: Sexual Attraction**
Illustration & Concept by Author

Figure 28: **A Spectrum of WORLDVIEWS: A Speculation**
Illustration & Concept by Author

CHAPTER 4: HUMANKIND

Figure 29: HUMAN DISPLACEMENT of Pre-humans 50-40,000 YBP
Illustrative Overlay of Out-migration Pathways: by Author.

Figure 30: The INFRASTRUCTURE of the INNER VOICE
Author's adaptation of image by Daniela Sammler et al, Dorsal and
Ventral Pathways for Prosody. *Current Biology* **25**, (2015)

Figure 31: The GLOBAL EXTINCTION of the genus *Homo*.
Author's re-conception of image from Julia Galway-Witham and
Chris Stringer. How did *Homo sapiens* evolve?
SCIENCE Vol **360** (2018)

Figure 32: NEANDERTHAL and HUMAN Body Cores
Credit: American Museum of Natural History, New York, NY
Graphics and Text and Author's adaptation of Body cores taken
from original AMNH images for comparative purposes.

Figure 33: COMPLEX PROJECTILES: Higher Velocity, Range, and Accuracy
Author's Freehand Sketch, reference from *Ethnoarchaeology*, Vol. **3**,
No. 2 (Fall 2011), p. 141, *Figure 1*: Photo. (Foreground)
Author's Freehand Sketch reference: Mark Denny, Atlatl Internal
Ballistics, *The Physics Teacher* **57**, 69(2019); (Background)
doi.org/10.1119/1.5088461

Figure 34: NEANDERTHAL ADAPTATION to 50°North Latitude
Graphics and Text by Author. Baseline Cranial Images from Gunz
et al. Neanderthal Introgression Sheds Light on Modern Human
Endocranial Globularity. 2019, *Current Biology* **29**, 120–127
(*Figure 1* on page 121). (Image: Philipp Gunz CC BY-NC-ND)
Note: the biological comparisons of Humans and Neanderthals
is central to the argument being put forth for the Theory of
Co·GENESIS. Related research papers, teams and work product
are fully attributed and credited under the provisions for Fair Use.
doi.org/10.1016/j.cub.2018.10.065

Figure 35: RNA DIFFERENTIATION of Neanderthals and Humans
Graphics and Text by Author. Images (upper and lower) from the
Natural History Museum at South Kensington, London, Science
News: nhm.ac.uk/discover/news/2021/february/modern-human-
origins-cannot-be-traced-back-to-a-single-point.html

Figure 36: DELIBERATIVE-CONSCIOUS Projections of the Human Mind
A Montage of Nine Illustrative Artifacts/Art, 40,000-30,000 YBP

- (A) The Lion Man. Stadel Cave, Germany, 40,000 YBP, Dagmar Hollmann/Wikimedia Commons "License: CC BY-SA 4.0"
- (B) Horses, Chauvet Cave, France. 35,000 YBP Image: Jean Clottes, Chauvet Cave Project.
- (C) Lions, Chauvet Cave, France. 35,000 YBP. Image: Jean Clottes, Chauvet Cave Project.
- (D) Bison, Chauvet Cave Replica, (2015). Image: Pierre Terdjman for The New York Times
- (E)The Griffon Vulture wing bone flute. Photo: H. Jensen. Copyright Universität Tübingen. Conard, N., Malina, M. & Münzel, S. New flutes document the earliest musical tradition in southwestern Germany. *Nature* **460**, (2009).
- (F) Lion Man Upper Body, Left Profile Photo: Dada629 CC-BY-SA-3.0 © Ulmer Museum
- (G) Lion Man Upper Body, Left Profile. Ink Drawing by Christina von Elm, ©Ulmer Museum
- (H) Female Figurine: *The Venus of Hohle Fels*, named after the cave of discovery. 40,000 YBP. Photo: H. Jensen
- (I) Female Figurine: *The Venus of Willendorf*, 30,000 YBP. Natural History Museum of Vienna, Austria. Photo by: Bjørn Christian Tørrissen / CC BY-SA 4.0

Figures 37A & 37B Hall of Bulls, Lascaux Cave, France, 18,000 YBP. Copyright 2015 Sisse Brimberg/ National Geographic (Publishing rights acquired).

CHAPTER 5: THE MECHANISMS

Figure 38: FIRST CELLULAR DIVISION after Fertilization
Illustration by Author: current DNA-centered Concept.

Figure 39: DUAL-SPINDLE FORMATION: Male and Female Legacy
Illustration by Author: Maintaining the Binary from one generation to the next. Compartmentalized male and female genes.

Figure 40: DNA UNARY vs. RNA BINARY and On-going Progression
Illustration by Author: Modern Synthesis (A) vs. Co·GENESIS (B)

Figure 41: CHARLES DARWIN (1809-1882).
Photo: "A photographic portrait of Darwin standing next to a pillar in the book Charles Darwin published in 1908, edited by his son Francis Darwin. Photo was taken c. 1881 by Messers. Elliot and Fry. Signature of Darwin is below the photograph. This media file is in the public domain in the United States." (Quote from Wiki/CC) Illustration by Ian Alexander CC BY-SA 4.0

Figure 42: PANGENESIS and Co·GENESIS: Information Flows.
Pangenesis Illustration by Ian Alexander CC BY-SA 4.0.
Co·GENESIS illustration by Author in the manner of Alexander.

Figure 43: NEMATODES: Earth's Most Populous Muti-cellular Animal.
Basic Diagram from *Florida Nematode Control Guide*; posted and cited on *Featured Creatures* Website ©University of Florida entnemdept.ufl.edu/creatures/nematode/soil_nematode.htm

Figure 44: miRNA: Molecular Messengers Generation-to-Generation
Image: miRNA stem loops for Humans and Nematodes
VTD-Own Work CC BY-SA 4.0

Figure 45: MALE LEGACY: Male miRNA
Graphics and Text by Author. Baseline Image © 2001 Benjamin Cummings, an Imprint of Addison Wesley Longman, Inc., Reference source: Marieb and Hoehn, ANATOMY and PHYSIOLOGY, Fifth Edition, Pearson Education, Inc. (2014), Chapter 26, The Reproductive System.

Figure 46: FEMALE LEGACY: Female miRNA
Image by Nicole Gross et al. MicroRNA Signaling in Embryo Development. *Biology* **6**, 34; *Figure* 3: "the potential to serve as a form of embryo-mother communication." © 2017 By the authors. Licensee MDPI, Basel, Switzerland. (CC BY) license: creativecommons.org/licenses/by/4.0/.

Figure 47: **RNA RECONCILIATION of the 2-cell to 16-Cell Embryo.**
Concept, Overlay Illustration, Graphics and Text by Author.
Baseline Images are modifications from *Gray's Anatomy*: reference
source: *Biology*. Provided by: OpenStax CNX. © Rice University
Located at: cnx.org/contents/185cbf87-c72e-48f5-b51e-f14f-
21b5eabd@10.8. License: CC BY: *Attribution*.

Figure 48: **STAGE ONE: Genetic/Body Demethalation Removing Parental Identities**
Co·GENESIS Concept and Illustration by Author

Figure 49: **STAGE TWO: Epigenetic/Mind Remethalation New Offspring Identity**
Co·GENESIS Concept and Illustration by Author

Figure 50: **RNA MIND/REPRODUCTION: Male and Female Pathways.**
Graphics and Text by Author. Image: NAU Biology Curriculum.
Website: 2.nau.edu/~gaud/bio301/content/hrmsex.htm. Cross-
reference to: University of Colorado, Integrative Physiology.

Figure 51: **RNA INITIATES the Male or Female Interpretation of the Body**
Graphics and Text by Author. Image from: Cao et al, Reproductive
role of miRNA in the hypothalamic-pituitary axis. *Molecular and
Cellular Neuroscience* **88**, pp. 130-137 (2018)
doi.org/10.1016/j.mcn.2018.01.008

CHAPTER 6: TRANSGENERATIONAL PATHWAYS

Figure 52: **RNA: Epigenetic/Genetic**
Separate pathways of Mind and Body
Concept and Illustration by Author.

Figure 53: **ENVIRONMENTAL ADAPTATION:**
RNA Imprinted Gene Flow.
Concept and Illustration by Author.

Figure 54: **RNA MIND: Applied, Anticipatory, and Pathways.**
Concept and Illustration by Author

Figure 55: **THE ORDER: Body/Heritable & Mind/Not Heritable.**
Concept and Illustration by Author

Figure 56: **28 Imprinted Male and Female Genes of Mind and Body.**
Supporting Graphics and Text by Author. Baseline Images, Text, and Diagram from: Olivia Ho-Shing, Catherine Dulac, Influences of genomic imprinting on brain function and behavior, Current *Opinion in Behavioral Sciences*, **25**, Pages 66-76, (2019) doi.org/10.1016/j.cobeha.2018.08.008.

Figure 57: **HYPOTHALAMUS: Master of Survival and Reproduction.**
Graphics, Text, and Overlay of Hypothalamus and Mapping of Male/Female Imprints and Intensity on Centerline of Brain Scan by Author. All Original Data and Research from: Olivia Ho-Shing, Catherine Dulac, Influences of genomic imprinting on brain function and behavior, *Current Opinion in Behavioral Sciences*, 25, Pages 66-76, (2019) doi.org/10.1016/j.cobeha.2018.08.008.

Figure 58: **The Nature of CONSCIOUSNESS.**
Graphics and Text by Author. Diagrams, Analysis, and Baseline Images from: Daniela Sammler et al., Dorsal and Ventral Pathways for Prosody. *Current Biology* 25, (2015)

Figure 59: **RNA NIGHTLY: Consolidation and Editing of Memory.**
Image is from *The Atlantic* article: Brain Cells Share Information with Virus-Like Capsules, by Ed Yong, January 12, 2018. The credit under the image is as follows:
"Virus-like shells budding off from one neuron and moving to another."
(Followed by the name of the illustrator) *Chris Manfre.*
The Atlantic article was on the work of Ellisa D. Pastuzyn et al.

Figure 60: **Co·GENESIS & Spontaneous Mutational Variation.**
Graphics and Text by Author. Baseline Images are from Houle et al, *Nature* **548** (7668) 2017: doi: 10.1038/nature23473.

Figure 61: **GENOMIC IMPRINTING: Social Evolution in Termites.**
Images are from: Matsuura, Kenji. "Genomic imprinting and evolution of insect societies." *Population Ecology* **62.1** (2020): 38-52. doi.org/10.1002/1438-390X.12026. Composition by Author.

Figure 62: **BEHAVIORAL INDIVIDUALITY | Separation of Body & Mind**
Linneweber et al., A neurodevelopmental origin of behavioral individuality in the Drosophila visual system. Science **6**; 367 6482) 1112-1119. (2020) doi: 10.1126/science.aaw 7182.
Reorganization/interpretation of *Science* Images by Author, Photo of Fly head with compound eyes: ALAMY, Inc., Credit: Razvan Comel Constantin/Alamy Stock Photo, (Publication Rights Acquired).

Figure 63 **The UNIVERSALITY of Co-GENESIS and RNA Primacy**
Overall Concept, Composition, Graphics and Text for the following Four Image Assembly: by Author.

- Image 1: Mosquito Body: Upper Left (UL)
  Illustration showing the anatomy of a mosquito (Culex pipiens)
  This work has been released into the public domain by its
  author, *LadyofHats*. This applies worldwide. (Wiki/CC)

- Image 2: Mosquito Brain/Neural Atlas/Female Mosquito (LL)
  Credit: Meg Younger/HHMI/Rockefeller University
  www.mosquitobrains.org/3d-rendering

- Image 3: Human Body (UR)
  Muscular System Picture, Anterior (Front) View, credited to
  Sport Fitness Advisor Staff, www.sport-fitness-advisor.com/
  muscular-system-picture.html

- Image 4: MRI of a Normal Brain.jpg.
  Source: Science Photo Library
  Credit: Fernando Da Cunha/SCIENCE PHOTO LIBRARY
  (Publication rights acquired).

CHAPTER 7: SIGNIFICANCE & MEANING

Figure 64: **ASEXUAL vs. SEXUAL: Reproduction & Adaptation**
Wilcox, Christie. Why Sex? Biologists Find New Explanations.
Quanta Magazine. (2020) Credit: Lucy Reading-Ikkanda/Quanta
www.quantamagazine.org/
why-sex-biologists-find-new-explanations-20200423/.

Figure 65: **SEPARATE MEMORY SYSTEMS: in Deliberation**
Cherry, K., The Parts of the brain: Biological Psychology, 2021,
VerywellMind, verywellmind.com/the-anatomy-of-the-brain-794895
Credit: Japanese Ministry of Education, Culture, Sports, and
Technology (MEXT) Integrated Database Project.

Figure 66: **BRAIN MAPPING: Track-Density Ratios of Twelve Individuals**
Choi Sang-Hanet et al., Track-Density Ratio Mapping with Fiber
Types in the Cerebral Cortex Using Diffusion-Weighted MRI.
Frontiers in Neuroanatomy **15**, (2021)
doi.10.3389/fnana.2021.715571

Figure 67: The UNIVERSAL RNA MIND: Humankind
NIH: Human Connectome Project (2011-2014)
www.humanconnectomeproject.org/category/news/ (click Gallery)
Plan View (lower image) Lecture: (Provided by Patric Hagmann,
CHUV-UNIL, Lausanne, Switzerland)
https://blog.myesr.org/mri-reveals-the-human-connectome/
Graphics and Text by Author

Figure 68: FOUNDATIONAL SCHOLARSHIP
Book Cover: Wilson, Edward O., *Consilience: The Unity of Knowledge*, Alfred A. Knopf, New York, N. Y., 1998,
Graphics and Text by Author

Figure 69: The RNA MIND (immortal) and The INDIVIDUAL (temporal)
Concept and Illustration by Author

Figure 70: INFORMATION FLOWS: RNA, DNA, and PROTEIN
Cobb M (2017) 60 years ago, Francis Crick changed the logic of
biology. *PLoS Biology* 15(9): e2003243.
doi.org/10.1371/journal.pbio.2003243
Concept and Illustration 69C by Author

Figure 71: Environmental, Neurological and Mind/Body Relationships
Three Concepts/Illustrations by Author

Figure 72: ILLEGAL HUMAN GENE EDITING: He Jiankui
Zang, Sarah. Chinese Scientists Are Outraged by Reports of Gene-
Edited Babies. *The Atlantic*, (2018)
Credit: Photo of He Jiankui by Mark Schiefelbein / AP
Credit: Image of Embryo selection/GettyImages.com/
(Publication rights acquired).

Figure 73: GENE EDITING and the Public's Perception
Book Cover: Isaacson, Walter: *The Code Breaker: Jennifer Doudna,
Gene Editing, and the Future of the Human Race*, Simon and
Schuster, New York, NY (2021).

Figure 74: INSTINCTIVE SUBCONSCIOUS vs. DELIBERATIVE CONSCIOUS
el-Showk, Sedeer. Neanderthal clues to brain evolution in humans.
Nature, **571**, S10-S11 (2019)
doi.org/10.1038/d41586-019-02210-6
© CC BY-NC-ND 4.0 (Image: Philipp Gunz) Note: the biological
comparisons of Humans and Neanderthals is central to the argument being put forth for the Theory of Co·GENESIS. Related
research papers, teams and work product are fully attributed and
credited under the provisions of Fair Use.

Figure 75: EXCEPTION and COMMONALITY within Mammals
- *Image 75A* Credit: Herculano-Houzel, Suzana. The remarkable, yet not extraordinary, human brain as a scaled-up primate brain and its associated cost. *PNAS*, **109** (supplement_1) (2012). (Fig.1, p. 10662) "Brain images are from the University of Wisconsin and Michigan State Comparative Mammalian Brain Collections" pnas.org/doi/epdf/10.1073/pnas.1201895109
- *Image 75B* Credit: ENCODE Project: Manolis Kellis , Barbara Wold, Michael P. Snyder, et al. Defining functional DNA elements in the human genome. *PNAS*, **111** (17) 2014 (Figure 1) doi.org/10.1073/pnas.1318948111

Figure 76: VALIDATION of ULTRACONSERVED ELEMENTS
Snetkova V, Ypsilanti AR, Akiyama JA, et al. Ultraconserved enhancer function does not require perfect sequence conservation. *Nature Genetics*. **53**(4):521-528. (2021) doi: 10.1038/s41588-021-00812-3.
Selected images: 76A, B, C and D are taken directly from the work of the research team, but have been Graphically composed and minor Text added by the Author for clarity and accessibility.

Figure 77: EVIDENCE of BRAIN and MIND IMPACTS
Snetkova V, Ypsilanti AR, Akiyama JA, et al. Ultraconserved enhancer function does not require perfect sequence conservation. *Nature Genetics*. **53**(4):521-528. (2021) doi: 10.1038/s41588-021-00812-3.
Selected images: 77A, B, C and D are taken directly from the work of the research team, but have been Graphically composed and minor Text added by the Author for clarity and accessibility.

Figure 78: BIOLOGICAL INTERACTION of MUSIC and MIND
Greenberg, D. M., Decety, J., & Gordon, I. The social neuroscience of music: Understanding the social brain through human song. *American Psychologist*, **76**(7), 1172-1185. (2021). doi: 10.1037/amp0000819
Image credit: Background artwork by Bryan Christie Design, Overlay design by Dr. David M. Greenberg
Graphic composition and text edit by Author.

AFTERWORD

Figure 79: The OBJECT as DEFINED by CARTESIAN GEOMETRY
By Jorge Stolfi - Own work, Public Domain, https://commons.wikimedia.org/w/index.php?curid=6692547 Wikimedia Commons.

ABOUT THE AUTHOR

Randolph Croxton founded Croxton Collaborative Architects, PC (CCA) in 1978—the firm's body of work includes the headquarters for the National Audubon Society, the Natural Resources Defense Council (NRDC) and the Environmental Defense Fund, in New York City as well as the U. S. Environmental Protection Agency (EPA) Headquarters in Washington, DC. These are among the CCA projects that first embodied the concept of architecture as a reconciliation of built and natural (biological) systems, and are recognized for establishing the principals of Green/Sustainable Architecture in the United States.[1] The health and well-being of the individual in this approach comes first. Croxton's contributions include the authorship of foundational advancements in the field:

The First U.S. Federal Grants for Green Architecture Strategy & Guidelines

- **Author**: January 1994—1st *NIST Grant* for the Lamont-Dougherty Earth Observatory Geochemistry Building for Columbia University, Palisades, NY
- **Author**: December 1994—2nd *NIST Grant* for expansion of the New England Aquarium, Boston, Massachusetts: "48 Metrics" of Sustainable/ Green Architecture.
- **Author**: 2004—*Sustainable Design Guidelines* for Redevelopment of the World Trade Center
  Public/Private Grant from Lower Manhattan Development Corporation (LMDC) and The Port Authority of New York and New Jersey (PANYNJ)

Published Books on Sustainability, Perception, and Consciousness

- **Co-Author**: May 1994 – *Audubon House* (John Wiley and Sons)
  "Audubon House illustrates a shining example of the future of architecture: a future where energy efficiency and the preservation of natural, historic, and human resources are the key..." —Amory B. Lovins, Director of Research, Rocky Mountain Institute
- **Author**: November 2015: *A Convergence of Two Minds: Origins of Self-Awareness and Identity* (Palustris Press)
  "Croxton argues that modern human minds succeed through the interaction of the distinctly male and female hemispheres of the brain…a thoughtful presentation of an intriguing theory. A thought-provoking explanation for the origins of personality." —Kirkus Reviews
- **Author**: December 2022: *Co-GENESIS—The RNA Mind: Perception and Consciousness* (Palustris Press Publication: April 2023)

Note: [1] *CCA received the 2005 National Leadership Award from the US Green Building Council for establishing the principles and practices of Green Architecture in America.*

www.ingramcontent.com/pod-product-compliance
Lightning Source LLC
Chambersburg PA
CBHW070439100426
42812CB00031B/3342/J